Advances in Stereotactic and Functional Neurosurgery 11

Proceedings of the 11th Meeting
of the European Society for Stereotactic
and Functional Neurosurgery,
Antalya 1994

Edited by
B. A. Meyerson and C. Ostertag

Acta Neurochirurgica
Supplement 64

Springer-Verlag Wien New York

Assoc. Professor Dr. Björn A. Meyerson
Department of Neurosurgery, Karolinska Hospital, Stockholm, Sweden

Professor Dr. Christoph Ostertag
Neurochirurgische Universitätsklinik, Freiburg, Federal Republic of Germany

Typesetting: Thomson Press, New Delhi, India

Printed on acid-free and chlorine free bleached paper

With 72 partly coloured Figures

ISSN 0065-1419 (Acta Neurochirurgica/Suppl.)
ISSN 0720-7972 (Advances in Stereotactic and Functional Neurosurgery)
ISBN-13: 978-3-7091-9421-8   e-ISBN-13: 978-3-7091-9419-5
DOI: 10.1007/978-3-7091-9419-5

Preface

This volume contains selected contributions from the XIth Meeting of the European Society for Stereotactic and Functional Neurosurgery held in September 1994 in Antalya/Turkey. Most of the papers deal with the many therapeutic and technical advancements made in this field of neurosurgery. The emergence of new stereotactic methodologies such as frameless stereotaxy and other forms of neuronavigation have become an indispensable tool for all types of neurosurgical operations. An increasing number of young neurosurgeons takes an interest in the neurosurgical approaches to the treatment of movement disorders, chronic pain and epilepsy. This is a clear sign of the growing awareness of the long neglected fact that these neurosurgical treatments can be offered to large patient populations.

Neurotransplantation as a novel treatment of Parkinson's disease has paved the way for the application of this technology for other indications. The pioneering work performed by the late Edward Hitchcock is reviewed here. There is a renewed interest in pallidotomy for dealing with certain forms of Parkinson's disease and certain aspects of this operation are discussed in another paper. Progress in the neurosurgical treatment of pain is dealt with by contributions on refined techniques of percutaneous cordotomy, DREZ operations and critical evaluations of spinal cord stimulation. A novel approach is a report on the experiences of treating cancer pain by intraspinal implantation of chromaffin cells. Several contributions cover the important issues of novel techniques for the study of neural dysfunction, peroperative monitoring with PET, microrecording, magneto-encephalography and other techniques.

This volume reflects the vitality of the field of stereotactic and functional neurosurgery and provides a broad review of current concepts and future developments of new treatment modalities.

Stockholm and Freiburg, October 1995

Björn A. Meyerson
Christoph Ostertag

Contents

Meyer, C.H.A., Detta, A., Kudoh, C.: Hitchcock's Experimental Series of Foetal Implants for Parkinson's Disease: Co-Grafting Ventral Mesencephalon and Striatum 1

Hirato, M., Ishihara, J., Horikoshi, S., Shibazaki, T., Ohye, C.: Parkinsonian Rigidity, Dopa-Induced Dyskinesia and Chorea – Dynamic Studies on the Basal Ganglia-Thalamocortical Motor Circuit Using PET Scan and Depth Microrecording 5

Dogali, M., Berić, A., Sterio, D., Eidelberg, D., Fazzini, E., Takikawa, S., Samelson, D. R., Devinsky, O., Kolodny, E. H.: Anatomic and Physiological Considerations in Pallidotomy for Parkinson's Disease ... 9

Whittle, I. R., Haddow, L. J.: CT Guided Thalamotomy for Movement Disorders in Multiple Sclerosis: Problems and Paradoxes 13

Mertens, P., Parise, M., Garcia-Larrea, L., Benneton, C., Millet, M. F., Sindou, M.: Long-Term Clinical, Electrophysiological and Urodynamic Effects of Chronic Intrathecal Baclofen Infusion for Treatment of Spinal Spasticity 17

Dones, I., Servello, D., Molteni, F., Mariani, G., Broggi, G.: A Neurophysiological Method for the Evaluation of Motor Performance in Spastic Walking Patients 26

Lozano, A. M., Hutchison, W. D., Dostrovsky, J. O.: Microelectrode Monitoring of Cortical and Subcortical Structures During Stereotactic Surgery 30

Broggi, G., Scaioli, V., Brock, S., Dones, I.: Neurophysiological Monitoring of Cranial Nerves During Posterior Fossa Surgery 35

Rousseau, J., Gibon, D., Coste, E., Blond, S., Pertuzon, B., Coche, B., Vasseur, C., Marchandise, X.: A Frameless Stereotaxic Localisation System Using MRI, CT and DSA 40

Taren, J., Ross, D., Lu, Y., Harmon, L.: 3D Laser Scanning for Image Guided Stereotactic Neurosurgery . 45

Doshi, P. K., Lemmieux, L., Fish, D. R., Shorvon, S. D., Harkness, W. H., Thomas, D. G. T.: Frameless Stereotaxy and Interactive Neurosurgery with the ISG Viewing Wand 49

Sandeman, D. R., Gill, S. S.: The Impact of Interactive Image Guided Surgery: The Bristol Experience with the ISG/Elekta Viewing Wand 54

Hellwig, D., Bauer, B. L., List-Hellwig, E.: Stereotactic Endoscopic Interventions in Cystic Brain Lesions 59

Kitchen, N.: Neurosurgery for Affective Disorders at Atkinson Morley's Hospital 1948–1994 64

Kuroda, R., Yorimae, A., Yamada, Y., Furuta, Y., Kim, A.: Frontal Cingulotomy Reconsidered from a WGA-HRP and c-Fos Study in Cat 69

Knutsson, E., Gransberg, L.: Localisation of Epileptic Foci with Multichannel Magnetoencephalography, MEG 74

Lehman, R. M., Kim, H.-I.: Partial Seizures with Onset in Central Area: Use of the Callosal Grid System for Localization 79

Sweet, W. H.: Pain – Old and New Methods of Study and Treatment 83

Kanpolat, Y., Caglar, S., Akyar, S., Temiz, C.: CT-Guided Pain Procedures for Intractable Pain in Malignancy 88

Zileli, M., Coşkun, E., Yegül, I., Uyar, M.: Electrophysiological Monitoring During CT-Guided Percutaneous Cordotomy 92

Lazorthes, Y., Bès, J. C., Sagen, J., Tafani, M., Tkaczuk, J., Sallerin, B., Nahri, I., Verdié, J. C., Ohayon, E., Caratero, C., Pappas, G. D.: Transplantation of Human Chromaffin Cells for Control of Intractable Cancer Pain 97

Linderoth, B., Gherardini, G., Ren, B., Lundeberg, T.: Severe Peripheral Ischemia After Vasospasm May Be Prevented by Spinal Cord Stimulation. A Preliminary Report of a Study in a Free-Flap Animal Model . . 101

North, R. B., Kidd, D. H., Piantadosi, S.: Spinal Cord Stimulation Versus Reoperation for Failed Back Surgery Syndrome: a Prospective, Randomized Study Design 106

Hassenbusch, S. J., Stanton-Hicks, M., Covington, E. C.: Spinal Cord Stimulation Versus Spinal Infusion for Low Back and Leg Pain 109

Fiume, D., Sherkat, S., Callovini, G. M., Parziale, G., Gazzeri, G.: Treatment of the Failed Back Surgery Syndrome Due to Lumbo-Sacral Epidural Fibrosis 116

Holsheimer, J., Barolat, G., Struijk, J. J., He, J.: Significance of the Spinal Cord Position in Spinal Cord Stimulation 119

Sindou, M., Chiha, M., Mertens, P.: Anatomical Findings in Microsurgical Vascular Decompression for Trigeminal Neuralgia. Correlations Between Topography of Pain and Site of the Neuro-Vascular Conflict 125

Gorecki, J. P., Nashold, B. S.: The Duke Experience with the Nucleus Caudalis DREZ Operation 128

Herregodts, P., Stadnik, T., De Ridder, F., D'Haens, J.: Cortical Stimulation for Central Neuropathic Pain: 3-D Surface MRI for Easy Determination of the Motor Cortex 132

Taira, T., Kawamura, H., Tanikawa, T., Kawabatake, H., Iseki, H., Ueda, A., Takakura, K.: A New Approach to the Control of Central Deafferentation Pain – Spinal Intrathecal Baclofen 136

Index of Keywords 139

Listed in Current Contents

Acta Neurochir (1995) [Suppl] 64: 1–4

Hitchcock's Experimental Series of Foetal Implants for Parkinson's Disease: Co-Grafting Ventral Mesencephalon and Striatum

C. H. A. Meyer, A. Detta, and **C. Kudoh**

Midland Centre for Neurosurgery and Neurology, Birmingham, U.K.

Summary

Following 4 previous experimental series of foetal implants (mesencephalon) to treat patients with Parkinson's disease subjects (N7) in the fifth series were treated with co-grafts of foetal mesencephalon and striatum implanted stereotactically into the caudate nucleus bilaterally. The clinical outcome, better than in the previous series, included improvements lasting through 18 months follow-up in activities of daily living, clinical neurological motor examination, timed motor tasks, and dyskinesia – with reduction in the patients' need for dopaminergic medication.

Keywords: Foetal transplants; Parkinsons's disease; stereotactic implants.

Introduction

When he died prematurely and unexpectedly E.R. Hitchcock, a pioneer in clinical neural transplantation, had performed foetal transplants to treat Parkinson's disease in 55 patients. Consecutive patients formed 5 series (Table 1) in which Hitchcock varied the site of implantation and the source of the implanted tissue, while aiming to keep other variables constant, as an evolving clinical experiment [2,3].

As discussed elsewhere [2,3] practices common to each implant procedure included: (a) patients with severe Parkinson's disease – at least stage 4, Hoehn and Yahr scale [4], in "off phases" – deteriorating despite optimal drug therapy; (b) graft tissue from single foetus, therapeutically aborted, relatively late gestational age of 11–20 weeks; (c) relatively long lag time, e.g. 5–12 hours, between foetal expulsion and dissection of graft en bloc; (d) graft injectate partially disaggregated into clumps (rather than cell suspen-sions) by mechanical agitation (no enzymatic dissociation); (e) stereotactic implantation under local anaesthesia through metal cannula – rather than open procedure; (f) no immunosuppressive therapy.

In post mortem examination of Parkinsonian patients who died (from causes unrelated to transplantation) more than 18 months after foetal implantation in Hitchcock's series I, II histological study showed that cells from second trimester foetuses could survive, mature, accumulate neuromelanin and express TH-IR when implanted in the striatum of patients who never had immunosuppression [1].

In line with our laboratory evidence for human second trimester tissue that co-culture with foetal striatum enhanced the survival of foetal mesencephalic neurones patients in series V were treated by co-grafts of foetal mesencephalon and striatum. When Hitchcock died seven patients described below had been treated in this way.

Methods

7 male patients, aged 46–62 (median 52) years, with Parkinson's disease, at least stage 4 or 5 on Hoehn and Yahr scale [4], having motor fluctuation ("on"/"off" phases) with dyskinesia and complying with CAPIT[4] criteria for neural transplantation, were treated by stereotactic intracerebral implantation of co-grafts of foetal mesencephalon and striatum.

For each patient graft material came from a single therapeutically-aborted foetus, gestational age 18–20 weeks. Foetal heart blood was tested to exclude HIV, Hepatitis A,B, Cytomegalovirus and Herpes Simplex infection. The aborted foetus was stored at 4 °C during 3–15 hours after expulsion before the dissection from it of blocks of foetal ventral mesencephalon and striatum. These were kept undissociated in Ham's F10 culture medium at 4 °C. Immediately before surgery the graft tissues were partially disaggregated by gentle agitation and the clumpy suspension, volume 0.5–0.8 ml, was injected stereotactically through bifrontal burr holes into each caudate head. In each case vital staining showed that viability of grafted cells was more than 55%.

Table 1. *Five Consecutive Experimental Series of Foetal Implants to Treat Patients with Parkinson's Disease.* Foetal transplantation – E.R. Hitchcock

Series	Year	Patients n	Patients Site of Implant	Foetus Source of graft
I	1988	12	R caudate	foetal mesenc.
II	1989	12	R putamen	foetal mesenc.
III	1990	12	R caudate	foetal mesenc.
IV	1991	12	Bilat caudate	foetal mesenc.
V	1992	7	Bilat caudate	foetal { mesenc. & striatum

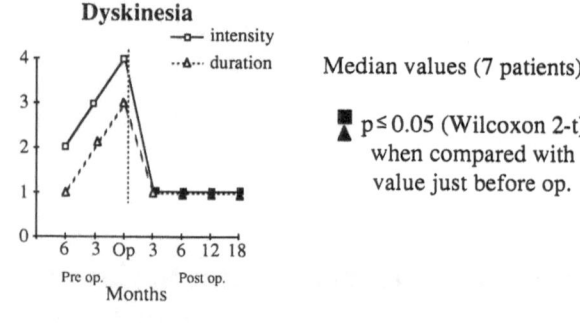

Median values (7 patients)

▪ p ≤ 0.05 (Wilcoxon 2-t)
▲ when compared with
    value just before op.

Established anticholinergic drug therapy was continued throughout the study. To simplify drug regimes DA receptor agonists and monoamine oxidase inhibitors were withdrawn at the outset and replaced with Laevodopa. Pre-operatively the Laevodopa medication was optimised to a dose giving the best therapeutic response [2]: post-operatively also the dosage was adjusted to give the best clinical neurological state.

Patients were evaluated 6 and 3 months and 1 week pre-operatively and 3,6,12,18 months post-operatively according to CAPIT protocols [4] including its rating scales for dyskinesia, modified Hoehn and Yahr scale for Parkinsonian severity, Unified Parkinson's Disease Rating Scale, and timed motor tasks. Other assessments: daily Laevodopa dosage; number of "on" hours per day; Northwest University Disability Scale [1] and Schwab and England Scale [5], both for activities of daily living.

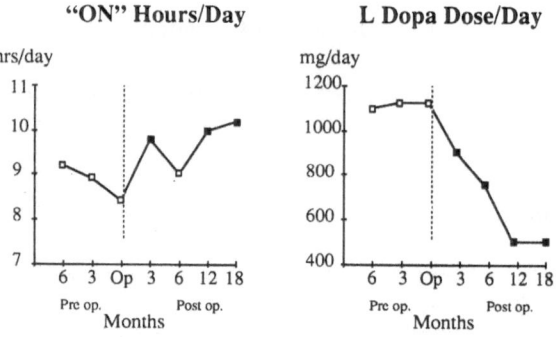

Fig. 1. Median values for group of Parkinsonian patients (N7). Laevodopa dosage, number of "on" hours per day, and dyskinesia according to CAPIT[4] rating scales for intensity (range 0–5) and duration per day (0–5). For dyskinesia high scores are bad

Results

Post-operatively for the 7 patients as a group there was substantial reduction in the optimal Laevodopa dosage (Fig. 1). For "off" phases when post-operative status is compared with the status immediately before surgery there was significant inprovement for all timed motor tasks (Fig. 2), ratings of disability for activities of daily living – NUDS, S&E, UPDRS (Fig. 3), clinical neurological motor examination (Fig. 3), intensity and duration of dyskinesia (Fig. 3), and the number of "on" hours per day (Fig. 1). For assessments during "on" phases (not listed here) post-operative improvement reached clinical significance at p 0.05 for a substantial, but smaller number of these parameters.

Discussion

In Hitchcock's first four series the foetal tissue came from ventral mesencephalon. In series I stereotactic right caudate implantation of foetal mesencephalon led to improvement in bradykinesia and rigidity, in activities of daily living, and in a timed task of repetitive pronation-supination of the forearm. Improvements were bilateral though voluntary movement benefited more on the side contralateral to the implant.

Following surgery the optimal dose of dopaminergic therapy could be reduced considerably. Dyskinesias were reduced. Improvements, more obvious during "off" phases, were most marked 6 months post-operatively and had waned by 12 months [2,3].

Clinical results were less good for series II (implantation into right putamen) than for series III (implantation, as in series I, into right caudate) so in series IV the foetal tissue, injected bilaterally, was placed in the caudate. Implanting grafts bilaterally did not greatly enhance the clinical results for the patients as a group though, as in earlier series, there were benefits for individual subjects.

Clinical outcome as shown here was very much better in series V when co-grafts of foetal mesencephalon and striatum were implanted into each of the right and left caudate heads. There were lasting benefits for all individual subjects. For the group as a whole, the Parkinsonian patients were much better post-operatively than pre-operatively throughout a range of parameters including the CAPIT assessments [4] widely used by contemporary investigators. Clinical benefits were sustained to the end of an 18 month follow-up in timed motor tasks, in activities of daily living, in the clinical motor state, in motor fluctuation (the relative

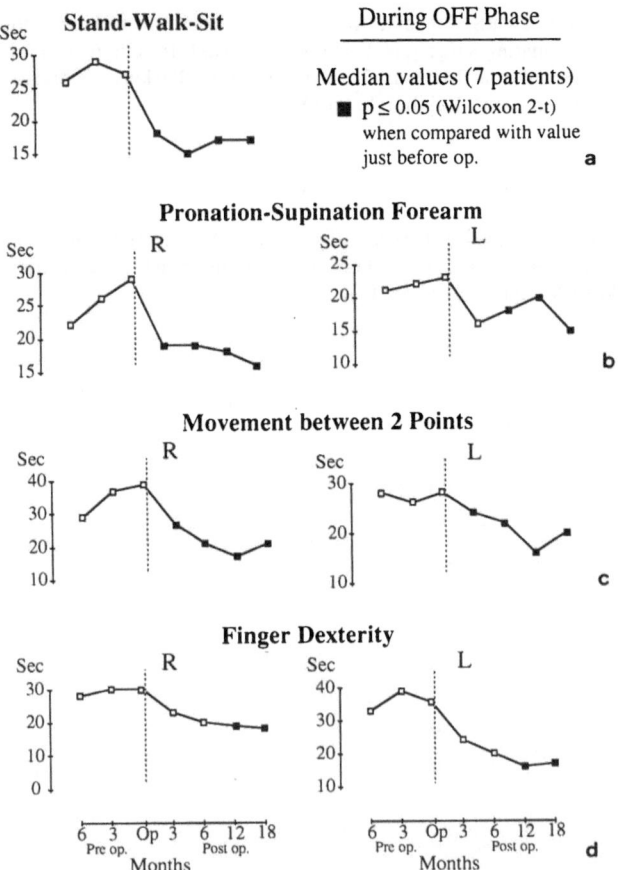

Fig. 2. Time to complete standardized motor tasks specified in CAPIT protocol [4]: (a) stand-walk-sit task; (b) pronation-supination of forearm, 20 repetitions; (c) movement of hand between points 30 cm apart, 10 to- and -fro repetitions; (d) finger dexterity: apposition of all four fingers successively to thumb, 10 repetitions of full cycle

time spent in the "on" state each day), and in dyskinesias. Post-operative improvement, in both "off" and "on" phases, reversed trends of progressive deterioration before surgery. The dosage of dopaminergic medication, optimised preoperatively, could be reduced greatly after surgery.

The much enhanced outcome for series V followed the co-grafting of foetal striatum with the foetal mesencephalon: in other important respects the procedure in series V adhered to the practices used in series I–IV – single foetus of 2nd trimester, graft preparation (mechanical partial disaggregation, injection as clumps), stereotactic instrumentation, no immunosuppression, management of antiparkinsonian drug therapy, patient selection.

The improved clinical results for series V support the use of these practices and indicate the value of co-grafting in treating Parkinson's disease by foetal intracerebral transplantation.

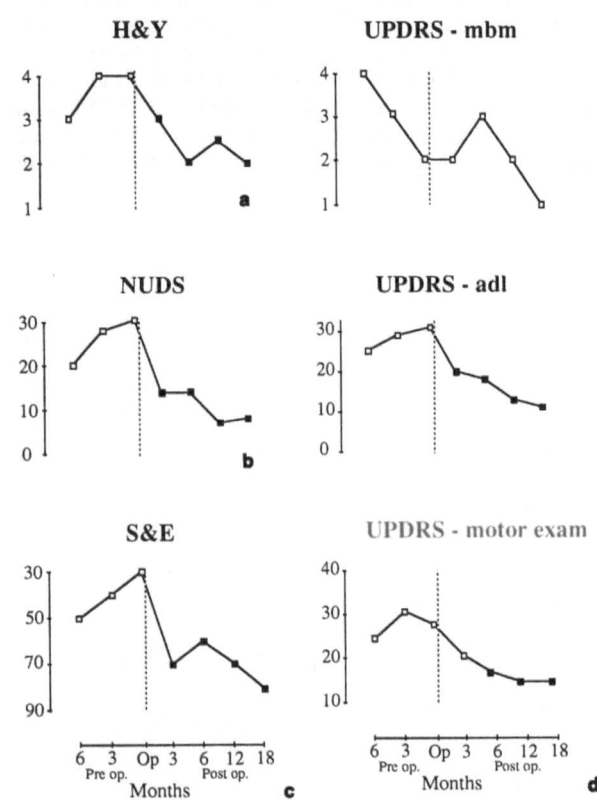

Fig. 3. Disability rating scales. Median values for group of Parkinsonian patients (N7) for assessments for (a) H&Y: Hoehn and Yahr scale, range 0–5, of severity of Parkinson's Disease [4]; (b) NUDS: North-West University Disability Scale [1] for activities of daily living, range 0–50; (c) S&E: Schwab and England [5] Disability scale for activities of daily living, range 0–100; (d) *UPDRS* Unified Parkinson's Disease Ratings Scale[4] with component scales MBM, mentation-behaviour-mood, score range 0–16; *adl* activities of daily living, range 0–52; *motor exam*, clinical neurological motor examination, range 0–108. For all scales direction upwards on the page is bad

Acknowledgement

Professor Hitchcock had valuable help from co-workers including B.T. Henderson, C.G. Clough, R.C. Hughes (Neurologists), B.G. Kenny (Neurosurgeon) and W. Mitchell (Neuro-technician); the Robert Nursing Home; Department of Cancer Studies at Birmingham University; the Blood Transfusion Service, Birmingham, The Parkinson's Disease Society. We thank Mrs. Cheryl Bickerton for typing this manuscript.

References

1. Bankiewicz KS, Whitwell HL, Sofroniew MV, *et al* (1993) Survival of TH-positive cells and graft indices post dopaminergic sprouting in patients with Parkinson's disease after intrastriatal grafting of foetal ventral mesencephalon. Soc Neurosci Abstr 864
2. Henderson BTH, Clough CG, Hughes RC, *et al* (1991) Implantation of human ventral mesencephalon to the right

caudate nucleus in advanced Parkinson's disease. Arch Neurol
48: 822–827

3. Hitchcock ER, Henderson BTH, Kenny BG, *et al* (1991)
Stereotactic implantation of foetal mesencephalon. In: Lindvall
O, Bjorklund A, Widner H (eds) Intracerebral transplantation,
in movement disorders. Elsevier, Amsterdam, pp 79–86

4. Langston JW, Widner H, Brooks D, *et al* (1991) Core assessment
programme for intracerebral transplantation (CAPIT) In:
Lindvall O, Bjorklund A, Widner H (eds) Intracerebral transplantation in movement disorders. Elsevier, Amsterdam, pp
227–241

5. Schwab RS, England AC, jr (1969) Projection technique for
evaluating surgery in Parkinson's disease. In: Gillingham FJ,
Donaldson IML (eds) 3rd Symposium on Parkinson's disease.
Livingstone, Edinburgh, pp 152–157

Correspondence: C.H.A. Meyer, FRACS, Midland Centre for
Neurosurgery and Neurology, Holly Lane, Smethwick, Warley,
West Midlands, B67 7JX, U.K.

Acta Neurochir (1995) [Suppl] 64: 5–8
© Springer-Verlag 1995

Parkinsonian Rigidity, Dopa-Induced Dyskinesia and Chorea – Dynamic Studies on the Basal Ganglia-Thalamocortical Motor Circuit Using PET Scan and Depth Microrecording

M. Hirato, J. Ishihara, S. Horikoshi, T. Shibazaki, and C. Ohye

Department of Neurosurgery, Gunma University School of Medicine, Gunma, Japan

Summary

Regional cerebral glucose metabolism (rCMRGlu-^{18}FDG) was measured in 6 cases with rigid type Parkinson's disease(PD) (2 cases with dopa-induced dyskinesia = DID), 6 cases with chorea(Ch), 5 cases with essential tremor (EssT) and 2 cases with normal subjects(N). The effects of L-Dopa on rCMRGlu was studied in 3 cases with PD. With the aid of depth microrecording study, stereotactic pallidotomy was performed in all cases with PD. Thalamotomy was performed in 3 cases with Ch. In the EssT and N group, the metabolic pattern was high in the frontal cortex (FCx) but low in the lenticular nucleus (LN). In contrast, all cases with a rigid type PD showed lower rCMRGlu in FCx(premotor, prefrontal area). However, 4 out of 6 cases were higher in LN than the control group. Administration of L-Dopa shifted rCMRGlu toward the normal pattern in this group. Five out of 6 cases with Ch represented higher rCMRGlu in FCx(3 focal, 2 diffuse) but lower in LN. Moreover, when DID occurred, it showed almost the same pattern as in Ch. Electrophysiological studies showed high background neuronal activity (BNA) in the medial segment of the globus pallidus(GP) but low BNA in the lateral segment of the GP in the rigid type of PD. In cases with Ch, irregular burst discharges were often encountered in ventro-oral thalamus. From these results, the on-going changes of basal ganglia-thalamocortical motor circuit in cases with a rigid type PD, DID and Ch are discussed. The underlying mechanisms of Parkinsonian rigidity was considered to contrast with those of DID and Ch within the same motor circuit.

Keywords: Parkinsonian rigidity; dopa-induced dyskinesia; chorea; basal ganglia-thalamocortical motor circuit.

Introduction

Based on clinical and experimental studies, both hypo- and hyperkinetic movement disorders may be accounted for respectively by specific disturbances within the basal ganglia-thalamocortical motor circuit [1–4]. However, data in living humans remains inconsistent. The present study was conducted to elucidate the underlying mechanisms of hypo-(parkinsonian rigidity) and hyperkinetic (dopa-induced dyskinesia, chorea) movement disorders in human beings. We performed dynamic studies on the basal ganglia-thalamocortical motor circuit using depth microrecording during the course of stereotactic pallidotomy or thalamotomy and positron emission tomography (PET) with ^{18}FDG and $C^{15}O_2$.

Patients and Methods

Quantitative estimation of regional cerebral glucose metabolism (rCMRGlu) was measured in 6 cases with the rigid type Parkinson's disease(PD) (2 cases with dopa-induced dyskinesia = DID), 6 cases with chorea(Ch), 5 cases with essential tremor(EssT) and 2 cases with normal subjects(N). The ratio of rCMRGlu in each structure (see below) to that in the ipsilateral cerebellar hemisphere was calculated, and the averaged ratio of each group was compared. We also studied the effect of L-Dopa on rCMRGlu in 3 cases with rigid type PD. PET studies were performed using ^{18}FDG with Sokoloff's method. Regional CMRGlu was measured in the thalamus, caudate nucleus, putamen, globus pallidus, and cerebral cortex. In our PET laboratory, the X-CT and the PET scanner are installed in parallel, and a single patient bed slides between them. Thus, the centers of each slice in X-CT and PET can be adjusted automatically, making precise anatomical comparisons possible. Furthermore, the cortical area was distinguished by the task dependent regional cerebral blood flow (rCBF) study using $C^{15}O_2$ in the same cases with FDG study. From our experience, sequential opposite finger movement caused an increase of rCBF on the contralateral cortical motor area.

With the aid of depth microrecording, stereotactic pallidotomy was performed in all cases with PD. Thalamotomy was performed in 3 cases with Ch. The details of the stereotactic operations with the aid of microrecording have been previously described. Before the operation, stereotactic MRI is done in order to get the necessary brain image from which the electrode track during the operation can be determined. Using Leksell's stereoractic apparatus, electrodes (usually applied in pairs) introduced through the burr hole pass through a selected plane on MRI. Thus, we can anatomically locate the electrode tip, in relation to the subcortical structure. Spontaneous electrical activity was continuously recorded when passing through the caudate nucleus toward the globus pallidus (first external and then internal segments) in cases with PD. In cases with Ch, the electrodes pass through the pallidal projection zone of the thalamus (VO) toward the ventralis intermedius. The signals are integrated on line (arbitrary unit), and, if necessary, a bar histogram can be made with the aid of a personal computer. Spontaneous discharge pattern and sensory responses to natural peripheral stimuli or to voluntary movement of the contralateral extremities can also be studied.

Results

PET Study

The FDG-PET image shows that a relative increase of rCMRGlu in the lenticular nucleus is marked in rigid type PD, while a relative increase of rCMRGlu in the frontal cortex is noticed in cases with EssT and in the normal control. In the former, the value of rCMRGlu and the average ratio of rCMRGlu were high in lenticular nucleus in 4 cases, but low in the frontal cortex in all cases. In EssT and in the normal control, values were high in the frontal cortex but low in the lenticular nucleus. Furthermore, in cases with a rigid type PD, administration of L-Dopa shifted the glucose metabolism toward a more normal pattern. In 3 cases in this group, rCBF was also studied during sequential opposite finger movements. As mentioned above, an zone of increased rCBF appeared in the contralateral cortical motor area. By superimposing the image on the FDG-PET Image representing the same slice, after administration of L-Dopa, rCMRGlu increased in the rostral portion of the cortical motor area of the frontal cortex, i.e. in the premotor or prefrontal area (Fig. 1). In cases with chorea, PET shows a relative decrease of rCMRGlu in lenticular nucleus (affected side) compared to EssT and to the normal control. However, in the frontal cortex, 5 out of 6 cases showed an increase of rCMRGlu (3 focal, 2 diffuse). In a case with chorea after an ischemic insult in the basal ganglia and with a concomitant history of PD, the value of rCMRGlu was low in the affected left lenticular nucleus but focally high in the frontal cortex (motor or premotor area) during the appearance of chorea. It decreased when the choreatic movements stopped. Furthermore, the DID group showed almost the same pattern as in Ch during dyskinetic movement.

Electrophysiological Study

In cases with a rigid type PD, electrophysiological studies revealed that the spontaneous activity of the

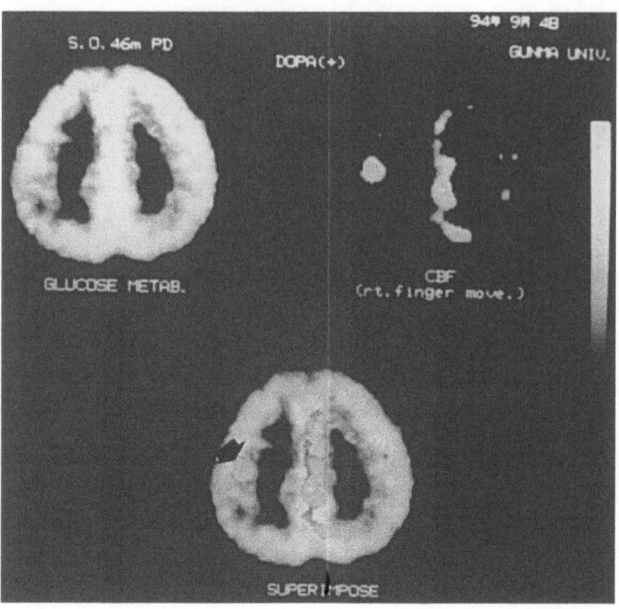

Fig. 1. Superimposition of the task rCBF-PET image on the FDG-PET image (case S.O.). Top left: FDG-PET image including the carebral cortex adjacent to the central sulcus after administration of L-Dopa. Top right: task rCBF-PET image during the sequential opposite finger movement of the right hand in the same case. This plane corresponds to that of the left FDG image. The increased zone of rCBF was found on the contralateral cortical motor area. Bottom: superimposition of the task rCBF-PET image on the FDG-PET image. Arrow head shows increased zone of rCBF on the cortical motor area. After administration of L-Dopa, rCMRGlu increases at the rostral portion of cortical motor area in the frontal cortex, that is the premotor or prefrontal area. The right side corresponds to the right side of the patient

caudate neurons was moderate, and that of GPe neurons relatively low, while there were vigorous discharges in the GPi. The trajectory through GPi was characterized by continuous high frequency spike discharges. Sometimes, sensory responses to natural peripheral stimuli or to voluntary movement in the contralateral extremities were noticed in the GPi (Fig. 2). However, in cases with chorea, irregular burst discharges were often encountered both in GPi (anatomically evaluated on MRI image) and in the pallidal projection zone of the thalamus (VO) (Fig. 3).

Discussion

In rigid type of PD, the glucose metabolism was relatively high in the lenticular nucleus but low in the frontal cortex (premotor or prefrontal area) compared to control cases. Administration of L-Dopa suggested that glucose metabolism shifted towards a normal pattern in the affected side. Electrophysiological

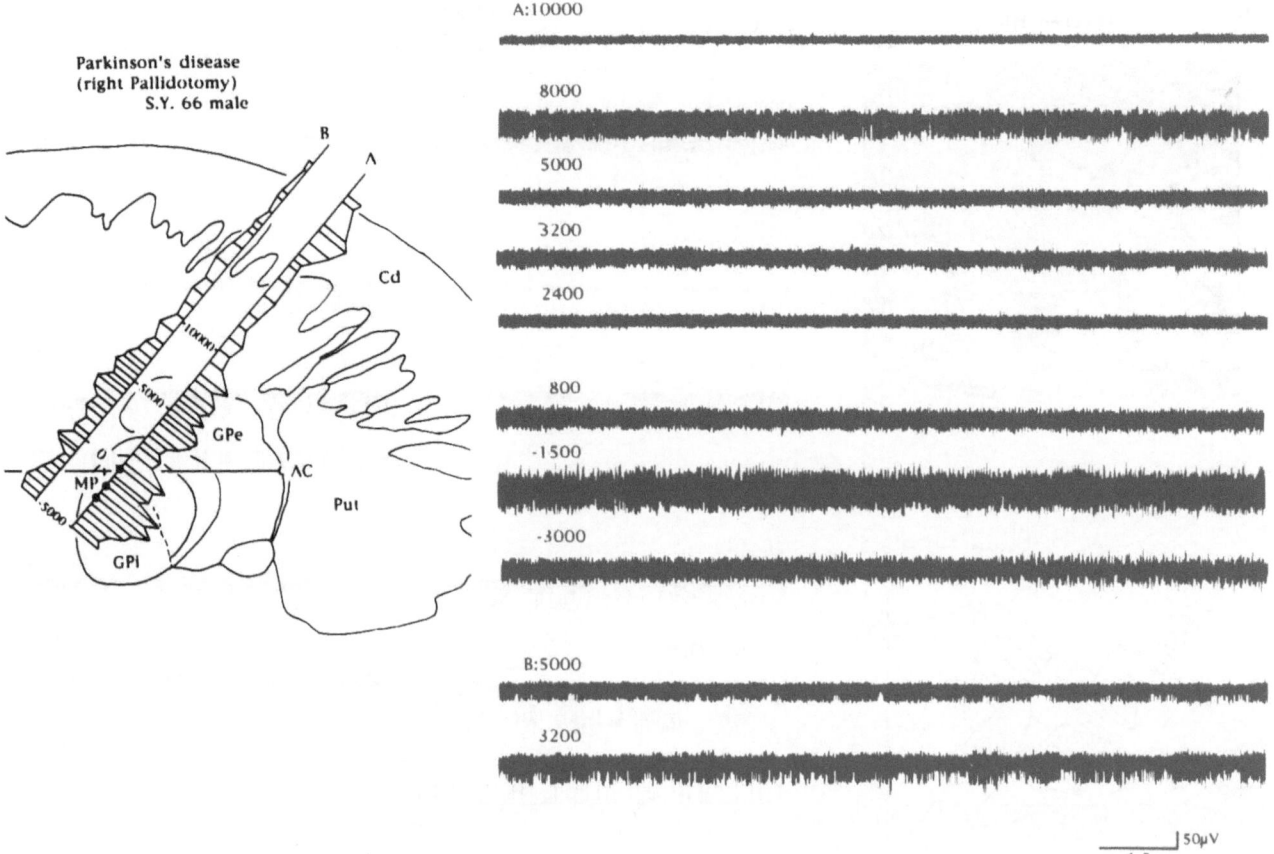

Fig. 2. A sequential change of background neural activity along the trajectories toward GPi recorded during a stereotactic pallidotomy (case S.Y., left rigid type PD). On the left is shown a bar histogram of the background neural activity on a sagittal view of the corresponding striatal zone. The pair of traces (A and B) represents the recording by a pair of electrodes set in parallel, rostrocaudally with 3 mm interspace. Black dots on the trajectory A show the location where sensory responses were obtained to peripheral natural stimulation. On the right, examples of the activity along the trajectories A and B at different levels are shown. Numbers indicate distance in microns from an arbitrary zero. Note the hyperactivity of GPi neurons. AC anterior commissure, MP midpoint of inter commissural line

recording revealed that the activity of GPe neurons was relatively low, while it was highly exaggerated in the GPi. Thus, FDG metabolism in the lenticular nucleus region was elevated in parallel with exaggerated spontaneous activity of GPi in the rigid type PD. These results are in agreement with changes previously observed in the MPTP-induced monkey model of Parkinson's disease [1,4], showing striatal dopamine deficiency due to nigrostriatal dysfunction. Furthermore, Parkinsonian rigidity may be much ameliorated by pallidotomy as well as by VO thalamotomy. Therefore, the present study supports and supplements previous findings and the recently proposed pathophysiological interpretations of the function of the striato-pallido-thalamocortical connections [1–4].

In contrast, Ch and DID displayed almost the same pattern, showing low glucose metabolism in the lenticular nucleus but high in the frontal cortex. In a case with chorea after an ischemic attack in the basal ganglia, the glucose metabolism was also low in the affected lenticular nucleus but focally high in the frontal cortex (motor, premotor area) during the appearance of chorea. It became low when the choreic movement stopped. Our electrophysiological study showed that irregular burst discharges were present both in the GPi and in the pallidal projection zone of the thalamus (VO). Therefore, DID and chorea must be mediated by the same motor circuit in rigidity [2,4]. However, the neural function of basal ganglia and of the frontal cortex in hyperkinesia is totally different from that in the rigid type PD.

These results suggest on-going changes of basal ganglia-thalamocortical motor circuit in cases with rigid type PD, DID and Ch. The underlying mechanisms of Parkinsonian rigidity appear to be different from those of DID and Ch within the same motor circuit.

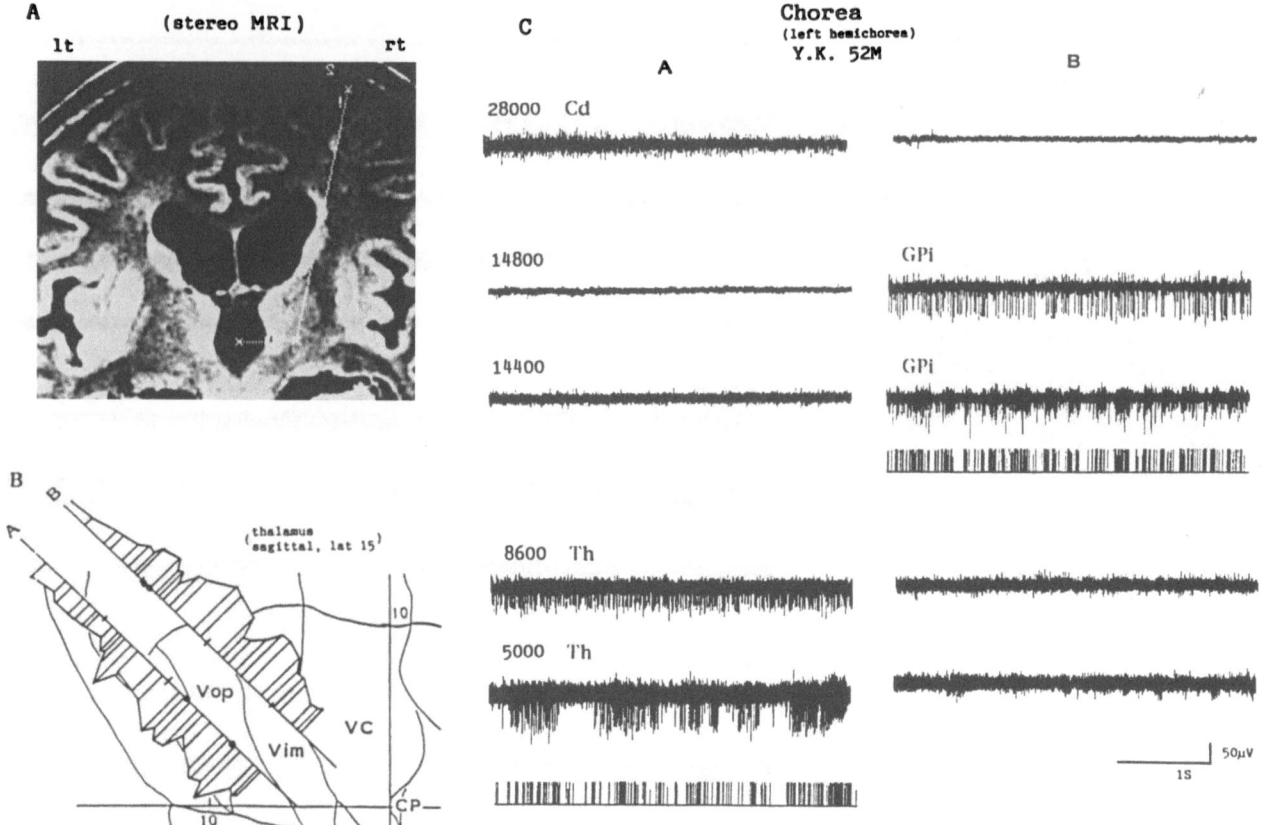

Fig. 3. A sequential change of background neural activity along the trajectories toward VIM in a stereotactic thalamotomy (case Y.K., left hemichorea). In the left top, coronal section of the thalamus parallel to the trajectory of the recording electrode. The oblique line shows a simulation of trajectory. In left bottom is shown a bar histogram of the background neural activity projected onto a sagittal view of the corresponding striato-thalamic zone. Black dots on the trajectory A and B show the location where irregular burst discharges were recorded as shown on the right. On the right, examples of the activity along the trajectory A and B at different levels are shown. Numbers indicate distance in microns from an arbitrary zero. Note that irregular burst discharges were recorded both in the GPi (trajectory B) and in the pallidal projection zone of the thalamus (VO) (trajectory A). *Vop* N. ventralis oralis posterius, *Vim* N. ventralis intermedius, *VC* N. ventralis caudalis, *CP* posterior commissure, 10 = 10mm

References

1. DeLong MR (1990) Primate models of movement disorders of basal ganglia origin. Trend Neurosci 13: 281–285
2. Hirato M, Ohye C *et al* (1993) Defferent cerebral metabolism between parkinsonian rigidity and hyperkinesia (DID, chorea, dystonia). A PET study. Adv Neurol 60: 511–514
3. Ohye C, Hirato M, *et al* (1994) Neural activity of the human basal ganglia in Parkinsonism compared to other motor disorders. In:

Percheron G *et al* (eds) Basal ganglia 4. Plenum, New York, pp 383–391
4. Page RD (1992) The use of thalamotomy in the treatment of levodopa-induced dyskinesia. Acta Neurochir (Wien) 114: 77–117

Correspondence: M. Hirato, Department of Neurosurgery, Gunma University School of Medicine, Gunma 371, Japan.

Acta Neurochir (1995) [Suppl] 64: 9–12

Anatomic and Physiological Considerations in Pallidotomy for Parkinson's Disease

M. Dogali[1], A. Berić[2], D. Sterio[2], D. Eidelberg[4], E. Fazzini[3], S. Takikawa[4], D. R. Samelson[1], O. Devinsky[3], and E. H. Kolodny[3]

Departments of [1]Neurosurgery, [2]Neurophysiology, [3]Neurology, NYU Medical Center-Hospital for Joint Diseases, New York, NY, and [4]Department of Neurology, North Shore University Hospital/Cornell University Medical Center, Manhasset, NY, U.S.A.

Summary

Our ongoing study of ventral pallidotomy for the control of Parkinson's disease in selected patients has provided the opportunity to explore the topographical and somatotopic organization of the human globus pallidus. Utilizing microelectrode techniques we have obtained recordings which were correlated with data from MPTP-parkinsonian primates. In addition, we performed pre- and post-operative FDG/PET scans in these patients. Our studies reveal similarities between the MPTP-parkinsonian primate model and human Parkinson's disease in terms of physiologic recordings and responses. However, we have encountered significant differences between dominant and non-dominant hemisphere representations, particularly for the hand, in the human. In addition, our PET studies confirmed, as in previous parkinsonian primate models, glucose hypermetabolism in the lenticular area of Parkinson's disease patients. This hypermetabolism is dramatically altered by creation of a lesion in the globus pallidus medialis. This is demonstrated by follow-up PET scans which reveal not only a decrease in metabolism of the operated lenticular region, but also in the frontal cortical projections. These combined observations of the cellular activity in the globus pallidus and the observed changes in PET metabolism support the selection of the pallidum for lesioning and control of Parkinson's disease, and offer insight into the underlying physiology of this disorder. The above physiological and PET data will be clinically correlated with our ongoing series of 35 + patients.

Keywords: Pallidotomy; hypermetabolism; microelectrode; PET.

Introduction

Surgical approaches to the globus pallidus for relief of symptoms arising from Parkinson's disease (PD) – rigidity, tremor, bradykinesia, and difficulty with gait and balance – date back to Meyers [1], who first reported results from division of the pallidofugal fibers at the ansa lenticularis via a transventricular approach in 1942. He later abandoned this for a subfrontal approach, with an electrode for thermocoagulation being passed through the anterior perforated substance and into the pallidum above the optic tract [2]. In the prestereotactic era, a number of other methodologies were developed, including the anterior choroidal lagation and chemopallidotomy techniques reported by Irving Cooper in 1953 [3] and 1958 [4].

A variety of stereotactic approaches to the globus pallidus (GP) were subsequently described by Spiegel and Wycis [5,6] Spiegel *et al.* [7], and many others [8–12]. While a significant improvement was often reported in PD patients, these various techniques did little to affect tremor and were accompanied by significant mortality and morbidity. In 1967, Lévy [13] presented a series of pallidotomies performed via a posteroventral approach, reporting successful relief of bradykinesia, rigidity and tremor without accompanying side effects and a low mortality.

With the advent of L-dopa therapy in the late 1960's, interest in surgical therapy for PD waned and relatively few pallidotomies were performed. In 1985, after reviewing a study published by Svennilson *et al.* [14] in 1960 whose methodology and results were similar to Levy's work, Laitinen *et al.* [15] resurrected the posteroventral pallidotomy technique for PD. Lesioning the posterior ventral GP in a series of patients, Laitinen *et al.* [15] reported significant improvement of parkinsonian symptoms.

The purpose of our study is twofold: 1) to refine the approach to this anatomical target, and 2) to elucidate the physiological events related to the GP in the parkinsonian state.

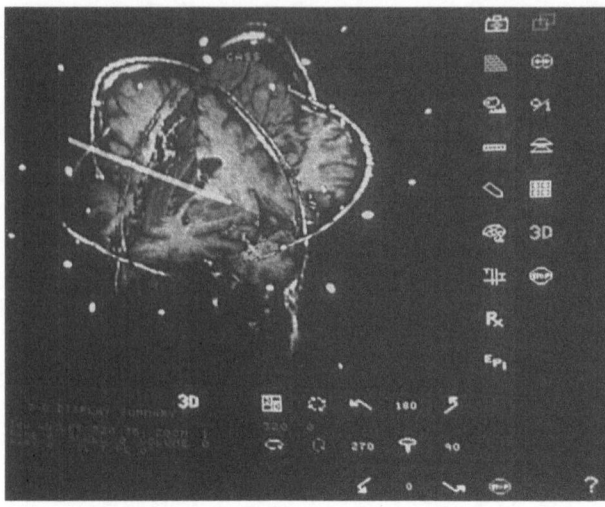

Fig. 1. Three-dimensional imaging with MRI

similar to those in DeLong's studies involving MPTP-induced parkinsonian primates [18,19]. During movements of contralateral limbs, the units consistently showed changes in discharge frequency in temporal relation to movement cycles. During ipsilateral limb movements, however, the discharge of the same unit showed no consistent relation to movements. A small number of units (2%) discharged nearly equally in relation to ipsilateral and contralateral limb movements. In five out of twelve righthanded patients we found finger movement responsive cells. Those cells were identified in four out of six left GP (13 finger responsive cells in six left GP), and one out of six right GP (only 1 finger responsive cell in six right GP). The difference is statistically significant ($\chi2 = 10.26$, p = 0.0035). This suggests possible hemispheric specialization with greater representation of the dominant hand [17]. Figure 5 shows the preoperative and six-month postoperative PET scans for a patient in our series. The postoperative PET scan demonstrates markedly decreased glucose metabolism on the surgical side in an area much greater than could be predicted by surgical volume. This expanded area is consistent with that reached by the frontal projections from the basal ganglia [20].

Conclusions

Both different frequency and different pattern of single cell activity were found in two pallidal segments: GPe and GPi. The discharge of 19% of cells in both pallidal segments was clearly modulated during passive movements of individual body parts. The cell clusters that alter discharge rates, a result of related movements, were identified. These findings suggest at least a partial somatotopic organization of the human GP and similarity with experimental results in both normal and MPTP-parkinsonian primates. If the concept that the GP is hyperactive in PD proves valid, identification of cells and their abnormalities would help in providing the optimal target for very localized lesions of the GP and result in predictable improvement of PD symptoms. These findings differ from those described in the MPTP-parkinsonian primate studies

Method

Figure 1 demonstrates the use of a GE Signa-1.5 Tesla MRI machine with data transferred to an independent workstation, allowing 3D reconstruction and confirmation of fiducials in space, thereby providing a more precise localization of the anatomical target [16]. MRI-obtained data allowed us to create computer reconstructions of GP topography in an attempt to relate its anatomy to its physiology. Figure 2 shows a computer-created view of the area of the GP [globus pallidus interna (GPi), globus pallidus externa (GPe), and the caudate nucleus]. As no data can currently define the precise optimal pallidal surgical site that yields maximal efficacy and minimal complications, we are developing neurophysiological maps of the GP based on intraoperative microelectrode recordings. Figure 3 shows different discharge patterns of cells registered in both GPe and GPi. Figure 4 shows the location of right GP cells responding to contralateral movement of different body parts in coronal (3 mm anterior from midcommissural line), axial (22 mm lateral from interhemispheric line) and sagittal (6 mm below intercommissural line) planes for six patients [17]. These findings are

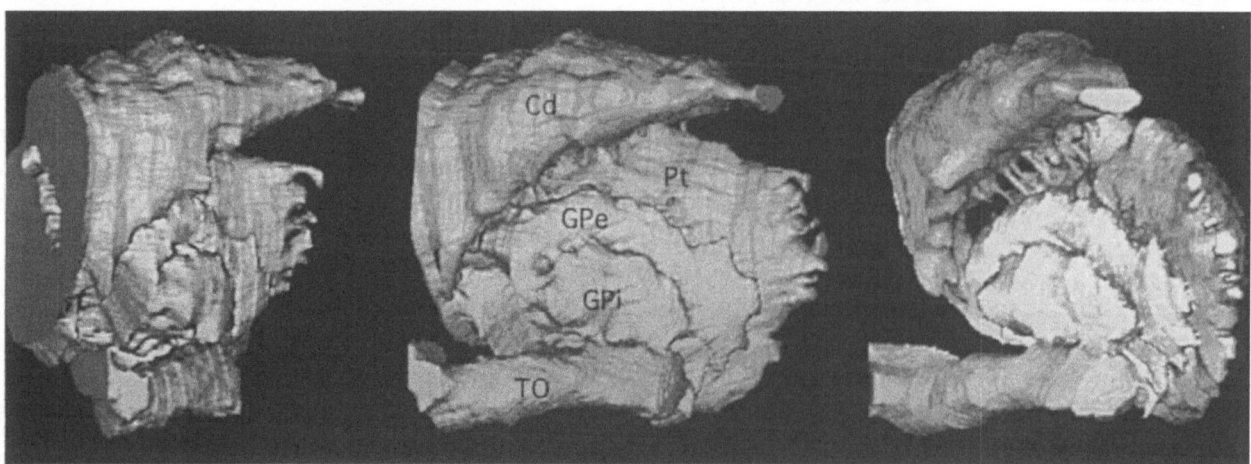

Fig. 2. Computer-created topography of right GP and adjacent structures

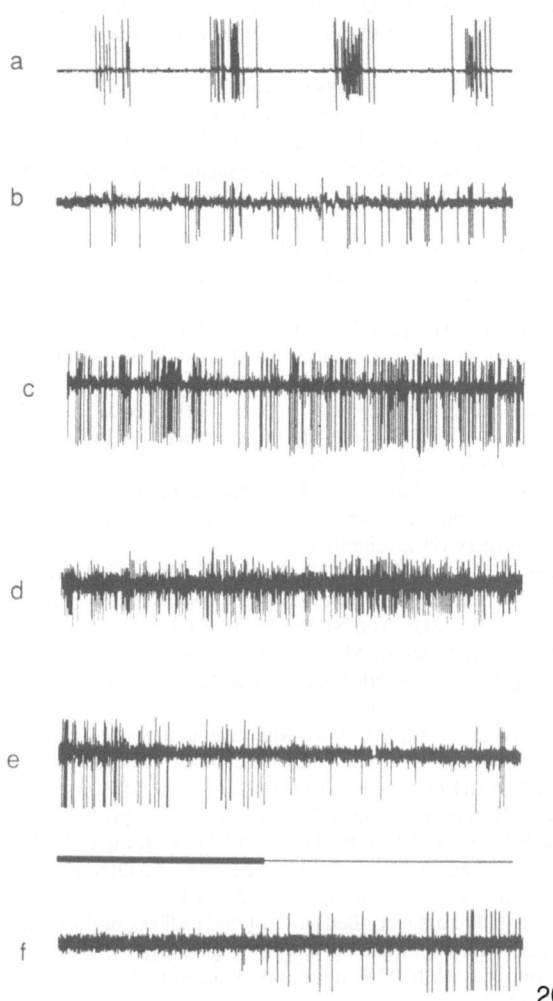

Fig. 3. Cell discharge of GPe and GPi. (a, b) Two types of cell discharges characteristic for GPe neurons (a) cell S-1538 = bursting activity; (b) cell B-2257 = low frequency, irregular activity); (c, d) high frequency, irregular discharges characteristic for GPi neurons; (e) activity of movement-related (F-7134) during hip and knee movements, activation during flexion and inhibition during extension; (f) activity of movement-related cell (F-8868, same patient, same trajectory, 1.7 mm below the cell F-7134) during hip and knee movements; inhibition during flexion and activation during extension. Thick horizontal line extension flexion, this horizontal line extension [17]

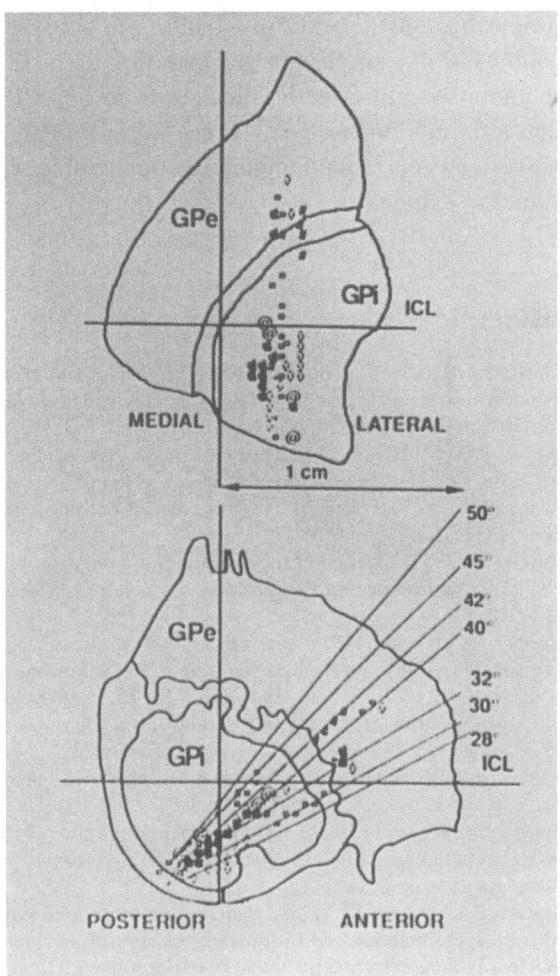

Fig. 4. Right GP cell response to contralateral body movement. *ICL* AC-PC intercommissural line

of DeLong *et al.* in the following ways. Differences between pallidal segments in PD patients were not as great as those reported in primates. This could reflect interspecies differences, but is most likely because our recordings were obtained in PD patients rather than healthy subjects. The difference in handedness is not reflected in any primate study. Both the clinical results and the neurophysiological data are, however, in keeping with one hypothetical basal ganglia circuitry of DeLong *et al.* modulating the distribution of

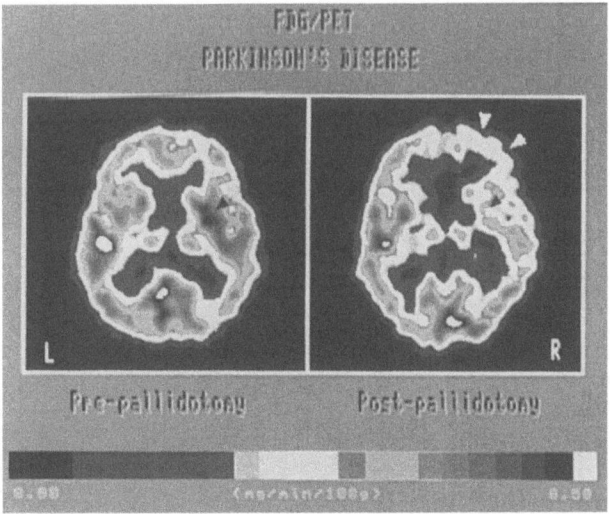

Fig. 5. Preoperative and postoperative PET

dopamine through the brain, thus confirming the selection of the GP as a surgical target. Selection of the GP is preliminarily supported by the data from our PET studies, although further study is needed in this area. Our investigations remain ongoing as our sample size continues to expand.

References

1. Meyers R (1942) Surgical interruption of the pallidofugal fibers: its effect on the syndrome of paralysis agitans and technical considerations in its application. NY St Med 42: 317–325
2. Meyers R (1958) Historical background and personal experiences in the surgical relieves of hyperkinesia and hypertonus. In: Fields, WS (ed) Pathogenesis and treatment of parkinsonism. Springfield, Thomas, pp 229–270
3. Copper IS (1953) Ligation of the anterior choroidal artery for involuntary movements of Parkinsonism. Psychiatr Q 1953; 27: 317–319
4. Cooper, IS, Bravo, GJ (1958) Anterior chroidal artery occlusion, chemopallidectomy and chemothalamectomy in Parkinsonism: a consecutive series of 700 operations. In: Fields WS (ed) Pathogenesis and treatment of Parkinsonism. Springfield, Thomas, pp 325–352
5. Spiegel EA, Wycis HT (1954) Ansotomy in paralysis agitans. Arch Neurol (Chicago) 71: 598–614
6. Spiegel EA, Wycis HT, Baird HW III (1958) Long-range effects of electropallidoansotomy in extrapyramidal and convulsive disorders. Neurology (Minneap) 8: 734–740
7. Spiegel EA, Wycis HT (1958) Pallido-ansotomy: anatomic-physiologic foundation and histopathologic control. In: Fields WS (ed) Pathogenesis and treatment of Parkinsonism. Thomas, Springfield, pp 86–105
8. Fénelon F, Thiébault (1950) Résultats du traitment neurochirurgical d'une rigidité parkinsoienne par intervention striopallidale unilatérale.Rev Neurol 83: 280
9. Guiot G, Brion S (1955) La chirugie pallidale dans les diskinésies. Sem Hôp (Paris) 31: 1838–1845
10. Narabayashi H, Okuma T (1955) Some contemplations on the role of the globus pallidus in parkinsonism. Brain Nerv 6: 157–161
11. Gildenberg PL (1960) Studies in stereoencephalotomy X. Variability of subcortical lesions produced by heating electrode and Cooper's balloon cannula. Confin Neurol 20: 53–65
12. Bertrand C, Martinez N (1959) Basal ganglia versus cortic-spinal tract lesions; their relative importance in the relief of tremor and rigidity. Rev Canad Biol 20: 365–375
13. Lévy A (1959) Die Pallidotomie beim Parkinsonsyndrom. Eine vergleichende anatomoradilogische Studie. Arch Psychiat Nervenkr 199: 487–507
14. Svennilson E, Torvik A, Lowe R, Leksell L (1960) Treatment of Parkinsonism by stereotactic thermolesions in the pallidal region. Acta Psychiat Scand 35: 358–377
15. Laitinen LV, Bergenheim T, Hariz MI: (1992) Leksell's posteroventral pallidotomy in the treatment of Parkinson's disease. J Neurosurg 76: 53–61
16. Dogali M, Roy R, Samelson DR (1995) Computer assisted MRI-guided targeting in functional and stereotactic neurosurgery. Neurosurgery 1995: submitted
17. Sterio D, Berić A, Dogali M, Fazzini E, Alfaro G, Devinsky O (1994) Neurophysiological properties of pallidal neurons in Parkinson's disease. Ann Neurol 35: 586–591
18. DeLong MR, Crutcher MD, Georgopoulos AP (1985) Primate globus pallidus and subthalamic nucleus: functional organization. J Neurophysiol 53: 530–543
19. Miller WC, DeLong MR (1987) Altered tonic activity of neurons in the globus pallidus and subthalamic nucleus in the primate MPTP model of parkinsonism. In: Carpenter MB, Jayarman A (eds) The basal ganglia II. Plenum, New York, pp 415–427
20. Dogali M, Fazzini E, Kolodny E, Eidelberg D, Sterio D, Devinsky O, Berić A (1995) Stereotactic ventral pallidotomy for Parkinson's disease. Neurology 45: 753–761
21. Filion M, Tremblay L, Bedard PJ (1991) Effects of dopamine agonists on the spontaneous activity of globus pallidus neurons in monkeys with MPTP-induced parkinsonism. Brain Res 547: 152–161

Correspondence: M. Dogali, M.D., Department of Neurological Surgery, University of California, Irvine, Medical Center, 101 The City Drive South, Building 3 – Route 81, Orange, CA 92668, U.S.A.

Acta Neurochir (1995) [Suppl] 64: 13–16

CT Guided Thalamotomy for Movement Disorders in Multiple Sclerosis: Problems and Paradoxes

I.R. Whittle and **L.J. Haddow**

Department of Clinical Neurosciences, Western General Hospital, Edinburgh, Scotland, U.K.

Summary

Unilateral ventrolateral (VL) thalamotomy for medically refractory tremorigenic movement disorders (MD) was performed in 9 patients with established multiple sclerosis. All patients had abolition of their coarse action/kinetic tremor with improvement in arm and hand function. In two patients some intention tremor either remained or was unmasked. Target coordinates ranged from 2 to -5 mm relative to the intercommissural line and from 8 to 16 mm lateral to the midline. There were no permanent surgical complications and the one stage peocedure under local anesthetic was well tolerated. Although there were also improvements in posture and speech in some patients the overall and longer term functional impact of surgery was, except in two patients, disappointing. Since multiple sclerosis is a spectrum of disease entities, and tremor may be only one manifestation of the disease, clinical studies that use comprehensive patient assessments and objective criteria may allow prediction of longer term functional outcome in specific patient subgroups. The specific aims of the stereotactic procedure in severely disabled patients with MS and MD must also be clear.

Keywords. Multiple sclerosis; tremor; thalamotomy; stereotactic surgery.

Indroduction

Stereotactic thalamotomy is an established treatment for disabling tremor in patients with multiple sclerosis (MS). Review of the literature suggests that in approximately 90% of patients there is immediate relief with 70% of patients remaining improved at one year [4–8]. Unfortunately, however, although there may be abolition of the tremor the functional significance of the results are unpredictable due to the multiplicity, activity and variable location of the lesions causing both the tremorigenic movement disorder (MD) and other clinical features of the multiple sclerosis [5–7]. Recent experience using CT image guided thalamotomy in a clinically heterogenous, but consecutive, series of MS patients with tremor is described to illustrate some of the problems and paradoxes in management of MS patients.

Methods

During the period from December 1989 to May 1994 12 patients with established and medically refractory tremorigenic MD were referred for surgery. These patients underwent pre-op video, clinical review and CT and MRI brain imaging. Indications for surgery were to abolish limb tremor with a view to improving the patient's functional status and/or facilitate general care. Ventrolateral (VL) thalamotomy was performed under local anesthesia (diazepam and droperidol), with dexamethasone cover. Where resting axial tremor was severe CT was performed using intravenous propofol. The BRW system was used with targets calculated from unreformatted axial CT images [10]. Lesions were made with a Radionics RF3B system. Intraoperative stimulations of the Gildenburg Stereotaxy Kit thermocouple probe (GSK-TC 1.3 mm radius, 4 mm tip length, of which 3 mm is insulated) were performed at 2–5 Hz and 50 Hz. Lesions were made 65 °C for 70 s. Postop video and clinical review were undertaken at 4 weeks and 3 months and one year.

Results

One 65 year old patient had a normal MR scan and was not considered to have MS. One 32 year old female had a slow (4 month) but steady transition in unilateral brachial tremor from a mixed kinetic/intention tremor to a fine intention tremor and operation was not recommended. One patient refused surgery. Nine patients, with variable disseminated neurological deficits due to multiple scerosis (Table 1), underwent unilateral VL thalamotomy. During pre-surgery hospitalization the kinetic tremor in one patient resolved considerably without any additional medication. Target coordinates straddled a large area of the VL complex (Fig. 1). Tremor was significantly attenuated following probe insertion in three patients. Intraoperative stimulations at the site of lesion making using 2 or 5 Hz produced

Table 1. *Profiles of the Nine Patients Undergoing Unilateral Stereotactic VL Thalamotomy for Tremor Due to MS*. All patients were unable to walk, dress, wash, groom or feed independently

Pt.	Age (yrs)	MS (yrs)	Tremor (mths)	Rest	Postural	Kinetic	Intention	Location	Barthel score	Comment
1	43	7	12	+	+	+ +	+	L > R (UL + LL)	4/20	spastic hemiparesis, INO, dysarthria
2	62	13	20	–	–	+	–	L (UL)	6/20	spastic hemiparesis dsyarthria, catheterised.
3	36	6	60	+	+	+ +	+ +	bilateral (UL)	8/20	ataxia, dysarthria, nystagmus
4	36	4	36	+	–	+ +	+ +	R > L (UL + LL)	8/20	spastic hemiparesis, ataxia
5	32	2	20	+	–	+ +	+ +	L > R (UL + LL)	9/20	spastic quadraparesis, nystagmus dysarthria
6	30	13	7	+	+	+ + +	+ +	R (UL + LL)	8/20	ataxia, mild spastic hemiparesis
7	35	17	20	+	+	+ +	+ +	bilateral (UL)	9/20	r spastic hemiparesis, INO dysarthria, nystagmus
8	34	3	18	+	+	+ + +	+ +	bilateral (UL + LL)	6/20	poor vision, dysarthria, spastic paraparesis, nystagmus
9	29	3	15	+	+	+ + +	+ +	bilateral (UL + LL)	7/20	spastic paraparesis, nystagmus

L left; *R* right; *UL* upper limb; *INO* internuclear ophthalmoplegia

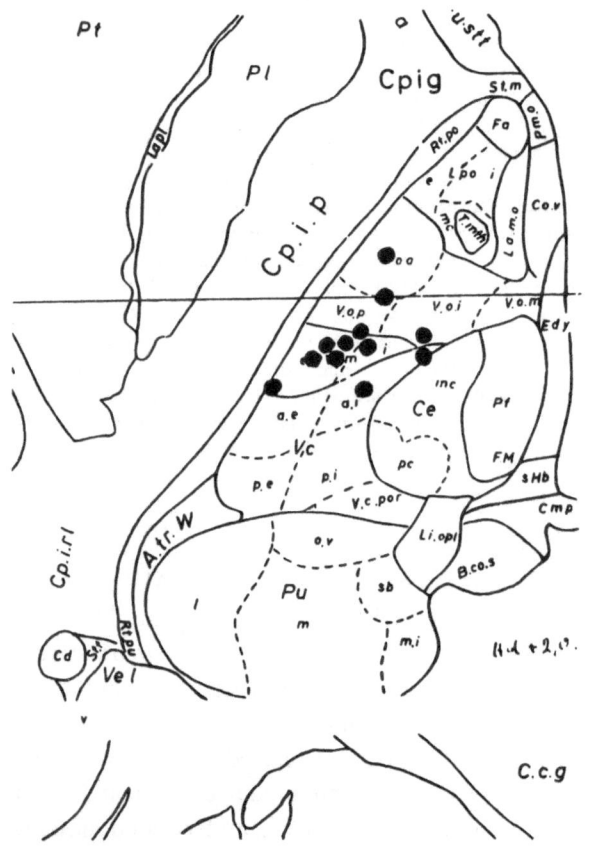

Fig. 1. Axial section of the thalamus 2 mm above the AC-PC plane showing the anatomical co-ordinates (black circles) at which the lesions were made in MS patients to abolish kinetic tremor. Although the lesions are spread over a large area most were in the region of the VL complex (*Voa, Vop, Vim*). Because of the variable width of the third ventricle in MS patients physiological assessment of target points is essential

either no phenomena (n = 6) or contralateral faciobrachial muscle twitching (n = 5). Stimulation at 50 Hz produced tetanic spasm in muscles of the contralateral upper limb (n = 5), nothing (n = 5), abolition of tremor (1) and focal contralateral dysaesthesia (1).

Course brachial action/kinetic tremor was abolished immediately in all patients with either a single (n = 7) or double (n = 2) Rf lesion with marked improvement in arm and hand function. In two patients a cerebeller intention tremor was unmasked following abolition of the kinetic tremor and in one patient intention tremor was unaltered by the lesion. In one of these patients the persisting intention tremor resolved over the 6 weeks following thalamotomy. In one patient passage of the probe into the VL complex almost completely abolished the kinetic tremor without an RF lesion being made (the Radionics lesion generator malfunctioned). This effect was maintained at three month follow up, despite an MR scan showing no lesion. As well as abolition of brachial and hand tremor with resultant improvement in arm and hand functions, other beneficial effects, particularly in the three months following surgery, were also noted on axial dystonia, speech volume, articulation and swallowing (Table 2). Transient psychomotor slowing and confusion (4–12 hours) occurred in five patients. Two patients had transient urinary dysfunction. Two patients became profoundly depressed after surgery.

Discussion

This study has confirmed previous reports that tremorigenic MD in patients with MS can be successfully treated by stereotactic thalamotomy with mini-

Table 2. *Clinical Outcome Following Unilateral Stereotactic Thalamotomy in Patients with Disabling Tremor Due to MS*. Although brachial tremor was abolished unilaterally in almost all patients, with dramatic improvement in arm function at six months postop only four patients were considered "better" by their neurologist. At one year postop only two patient (Cases 4.5) were still significantly better

	Tremor		
Pt.	Kinetic	Intention	Comment
1	abolished	reduced	excellent short term improvement of arm function, only 4/12 postop
2	improved	none	arm too spastic and ataxic for useful function.
3	abolished	unmasked	residual intention tremor resolved over 6/52 post-op, remains profoundly ataxic.
4	improved	improved	resting tremor also improved, mild incontinence post-op.
5	abolished	abolished	resting tremor (head) also improved, swallowing better, mild, transient post-op incontinence of urine.
6	abolished	abolished	postural tremor also abolished, major relapse of MS 4/12 postop.
7	abolished	reduced	dysarthria less, truncal dystonia less, postop depression; effects due to "microthalamotomy" due to probe passage.
8	abolished	minimal	dysarthria less, truncal dystonia decreased, post-op depression
9	abolished	minimal	postural tremor improved also.

mal surgical morbidity [4,6,7]. From the technical viewpoint short acting general anesthetic agents such as propofol are extremely useful for patients with severe axial resting tremor. Use of propofol can facilitate fixation of the stereotactic base ring and also optimise CT scan quality in these patients. Selection of co-ordinates for target points should centre on the VL complex but physiological testing will be required in most patients to optimise lesion placement. The characteristic physiological responses seen following VL stimulation in non-MS patients were often not reproduced in this cohort. This finding may be related to the focal and diffuse atrophy or damage of the brain that is characteristic of MS. The neuropathological process may also lead to gross widening of the third ventricle, atypical thalamic nuclear relationships to conventional stereotactic reference points and abnormal discharge patterns from the nucleus ventro-intermedius (Vim) region [1,7].

Action or kinetic tremor, which in MS patients is often called "rubral" tremor, usually responds very well to an appropriately placed radiofrequency lesion, however intention tremor seems much more refractory. Indeed, in some patients it was unmasked by removal of the "rubral" tremor. Residual, unmasked or later recurrence of intention tremor was one impediment to significant functional improvement in most of our cases. The other major problem in this patient cohort is the diffuse nature of neurological disease. The two patients who had a sustained functional improvement following thalamotomy had the least systemic neurological disability and were therefore able to use their tremor free limbs well. Most of the other patients were severely disabled with spasticity, brain stem and cerebellar disease as well as multilimb and axial tremor. These problems are uneffected by unilateral abolition of "rubral" tremor. Additional beneficial effects in these patients could be considered easing of their disabled condition by improving voice volume, articulation and resting tremor and facilitated nursing care. Nonetheless two patients became profoundly depressed, despite abolition of tremor, presumably because of their frustration with their residual disabilities. Such negative feelings have previously been noted [8]. Functionally most patients remained considerably handicapped [6,8]. Realistic goals are therefore essential when selecting patients for surgery [5].

In some patients with severe bilateral disabling tremor it is theoretically attractive to consider a thalamotomy on one side and the implantation of a thalamic stimulator in the contralateral thalamus [3, 7]. The latter procedure also has the benefit of not causing additional lesions in patients who already have a disseminated disease. Such an integrated approach could conceivably lead to a better functional outcome. If such a study was to performed rigorous and objective assessments of both the tremor [2] and functional status of the patient should be undertaken. Unfortunately there are major shortcomings and difficulties with many indices such as the Kurtzke Rating scale, the Rankin and Bartel scales [6,9].

Although all patients were considered to have stable neurological disease when referred for surgery two underwent spontaneous changes in the composition of the brachial tremor. Furthermore one patient had sustained abolition of tremor merely after passage of the stimulating electrode into the VL complex. Post-operative MR in this patient field to reveal any lesion. The physiological bases of these phenomenon are unclear but presumably relate to the balance between demyelination and oligodendroglial function in the inflammatory plaques. Another unusual feature of the series is the relative lack of surgical side effects. The relative safety of VL thalamotomy in MS patients varies with some having few problems [4,6,7] whilst other have major difficulties [5,8]. Given the multiplicity and bilateral nature of the MR visualized plaques in most patients the absence of post-thalamotomy dysphonia, dysphasia, hemiparesis or dysphagia was surprising.

Acknowledgement

This work was partly funded by a Scottish Office Home and Health Department Grant to LJH.

References

1. Andrew J, Rice-Edwards M, Rudolf N (1974) The placement of steroetactic lesions for involuntary movement disorders other than Parkinson's Disease. Acta Neurochir [Suppl] 21: 39–47

2. Bain P (1993) A combined clinical and neurophysiological approach to the study of patients with tremor. J Neurol Neurosurg Psychiatry 69: 839–844

3. Benabid AL, Pollack P, Seigneuret E, *et al* (1993) Chronic Vim stimulation in Parkinson's disease, essential tremor and extrapyramidal dyskinesias. Acta Neurochir [Suppl] 58: 39–44

4. Cooper IS (1967) Relief of intention tremor of multiple sclerosis by thalamic surgery. JAMA 199: 99–104

5. Hauptvogel H, Poser S, Orthner H, *et al* (1975) Indikationen zur stereotaktischen Operation bei Patienten mit multipler Sklerose. J Neurol 210: 239–251

6. Hitchcock ER, Flint GA, Gutowski NJ (1987) Thalamotomy for movement disoders; a critical appraisal. Acta Neurochir [Suppl] 39: 61–65

7. Siegfried J (1993) Therapeutic stereotactic procedures on the thalamus for motor movement disorders. Acta Neurochir (Wien) 124: 14–18

8. Speelman JD, van Manen J (1984) Stereotactic thalamotomy for the relief of intention tremor of multiple sclerosis. J Neurol Neurosurg Psychiatry 47: 596–599

9. van Gijn J (1992) Measurement of outcome in stroke prevention trials. Cerebrovasc Dis 2 [Suppl 1]: 23–34

10. Whittle IR, O'Sullivan MG, Ironside JW, *et al* (1993) Accuracy of ventrolateral thalamic nucleus localization using unreformatted CT scans and the BRW system: Experimental studies and clinical findings during functional neurosurgery. Acta Neurochir [Suppl] 58: 61–64

Correspondence: Mr. I.R. Whittle MD, PhD, FRACS, FRCSE(SN), Western General Hospital, Edinburgh EH4 2XU, Scotland, U.K.

Acta Neurochir (1995) [Suppl] 64: 17–25

Long-Term Clinical, Electrophysiological and Urodynamic Effects of Chronic Intrathecal Baclofen Infusion for Treatment of Spinal Spasticity

P. Mertens[1], M. Parise[1], L. Garcia-Larrea[2], C. Benneton[3], M. F. Millet[3], and M. Sindou[1]

[1] Department of Neurosurgery, [2]Department of Clinical Neurophysiology, and [3]Department of Rehabilitation, Hopital Neurologique, Lyon, France

Summary

Seventeen patients with severe disabling spinal spasticity were selected and treated by chronic intrathecal baclofen infusion using an implanted programmable pump. Nine patients were tetraparetic, seven were paraplegic and one paraparetic. Patients were regularly followed for 5 to 69 months (mean 37,5 months).

The *clinical efficacy* of baclofen was estimated by means of evaluation of: hypertonia, spasms, pain and functional disability. All patients experienced significant amelioration of quality of life secondary to reduction of hypertonia, spasms and pain related to contractures. Neurogenic pain improved in 3 cases and remained unchanged in 3 others. In patients whose motor functions were partially preserved, various degrees of motor improvement were detected.

Electrophysiological recordings of *Polysynaptic flexion reflexes* (FR) were obtained to control conditions, and under intrathecal baclofen, in order to quantify the spinal excitability responsible for spontaneous or induced spasms. Flexion reflex threshold was increased and amplitude proved to be very significantly reduced by chronic baclofen infusion in all our patients.

Twelve patients with *neurogenic bladder dysfunction* were also evaluated by a clinically oriented questionnaire and by quantitative urodynamic recordings, before and after pump implantation. In patients with normal micturition, this was not changed by intrathecal baclofen. In patients with spastic bladder, intrathecal baclofen produced a decrease of detrusor hypertonia and hyperactivity in 50% of cases, with reduction of leakage and increase in functional bladder capacity.

Keywords: Spasticity; neurogenic bladder, intrathecal baclofen; flexion reflex.

Introduction

Chronic infusion of baclofen into the subarachnoid space was introduced in 1984 by Penn and Kroin [10,19,20]. Baclofen, is a GABA-B receptor agonist, with pre-synaptic [3] and possibly post-synaptic [1] inhibitory actions in superficial layers of the spinal cord dorsal gray matter. High concentrations of this drug at these specific receptor sites permit a marked reduction of the spasms and rigidity by decreasing motor neuron excitability [10,16]. In recent years, technological advances have permitted a safer administration of intrathecal baclofen using the implanted programmable pumps.

In this article, we report a 17 patients series with severe chronic spasticity treated by intrathecal baclofen and followed from 5 to 69 months (37.5 months on average). Polysynaptic flexion reflexes (FR) were recorded first in control conditions and then under intrathecal baclofen, in order to quantify the spinal excitability responsible for spontaneous or induced spasms. Assessments obtained by clinical spasm scales were compared with polysynaptic flexion reflexes recordings. The baclofen effects on lower urinary functions were also analysed by urodynamic studies in twelve patients with neurogenic bladder dysfunction.

Material and Methods

I. Patients Selection

Twenty four patients were selected for an initial trial with intrathecal baclofen on the following criteria: 1) Severe, chronic and disabling spasticity due to a spinal lesion; 2) Spasticity refractory to oral drugs (including oral Baclofen), or unacceptable side-effects with effective doses; 3) Absence of hypersensitivity (allergy) to baclofen; 4) Favorable environment for rigorous and regular outpatient follow-up; 5) Adult age (between 18 and 65 years), and cognitive capacity to give informed consent.

Table 1. *Functional Disability Score for Spastic Paraplegic Patients*

Pain	0	No pain
	1	Unfrequent pain and or minor intensity; the patient does not complain spontaneously; without any consequence on daily life
	2	Frequent pain and or moderate intensity; the patient complain spontaneously; but without any practical consequence on daily life
	3	Very frequent pain and/or severe intensity; with practical consequence in daily life
	4	Permanent and unbearable pain
Spasms	0	No spasms
	1	Unfrequent spasms and of minor intensity; only provoked by mobilization; without any functional consequenses
	2	Frequent spasms and or moderate intensity; occurring at mobilization and also spontaneously; but without impairing the patients confort in sitting and lying position
	3	Very frequent spasms and/or severe intensity; impairing the sitting and lying position (awaking patient a night)
	4	Almost constants spasms, making impossible any correct sitting position and even lying position
Wheel chair	0	No disconfort
	1	Minor disconfort, not reducing significantly sitting time in weelchair
	2	Moderate disconfort, reducing sitting time in weelchair
	3	Strapping is required to mantain correct sitting position
	4	Weelchair impossible
Transfers	0	Easy, alone
	1	Possible alone, but difficult
	2	Need for one person helping
	3	Possible but difficult with one person helping
	4	Need for two persons helping
Washing and dressing	0	Easy, alone
	1	Possible alone, but difficult
	2	Need for one person helping
	3	Possible but difficult with one person helping
	4	Need for two persons helping

II. Acute Intrathecal Baclofen Trial

In this initial trial, baclofen was injected via a lumbar puncture or a subcutaneous port connected to a catheter in the lumbar subarachnoid space. The initial dose (5 to 25 µg) was increased by steps until a satisfactory response was obtained or maximal dose of 100 µg was reached.

The clinical efficacy of acute baclofen was estimated by means of a composite evaluation which included the Ashworth, the spasm, and the functional scales (Table 1), as well as standardized assessment of walking abilities and amplitude of articular movements.

The acute trial was considered satisfactory if either a two-point reduction in the Ashworth and spasms scores, or a decrease of the functional score to values lower than 10/20, or both, were obtained. However, in patients who were able to walk before the baclofen trial, reduction of voluntary motricity due to baclofen-induced hypotonia was an absolute exclusion criteria even if the rest of the evaluation proved satisfactory.

Based on these evaluations, 17 out of the 24 patients were considered as good candidates for pump implantation. The reasons for exclusion of the remainning 7 patients were, in five cases a significant reduction in the walking performances owing to baclofen-induced excessive hypotonia, even at low doses (5 µg), while in two other cases even a dose of 100 µg was unable to decrease the pathological hypertonia.

III. Patients Undergoing to Chronic Baclofen Infusion

The 17 patients in whom a programmable pump was implanted were regularly followed for 5 to 69 months (mean 37.5 months). The age, sex, diagnosis and clinical conditions are summarized in Table 2. One single patient was able to stand up and walk alone for several meters, and another could do so with the help of crutches. All other patients were wheelchair dependent or bedridden, even if in three cases motor function was not completely abolished.

Eleven patients had pain related to spasms or central neurogenic pain. Neurological bladder dysfunction was present in 14 patients. One patient had a cystectomy performed due to severe abdominal injury.

IV. The Implanted Pumps

In all cases, the pump used was a Synchromed Programmable Pump (Model 8615-Medtronic Inc, Minneapolis-USA). The surgical technic has been described in detail elsewhere [12, 18–20].

IV. Post-Implantation Management

During the post-operative period, the pump was programmed to deliver a total daily dose approximately twice the minimal efficient dose in the acute test. Oral antispastic medication was tapered gradually.

The patients were reviewed regularly every three months. At the moment of pump refilling the patients status was evaluated with the same quantitative scales used during the acute trial (Tables 1). The dose and infusion mode were adjusted, if necessary, to maintain a good clinical and functional condition (Ashworth and spasm scores < 3 and functional score < 10).

V. Flexion Reflex Measurements

Flexion-reflexes (FR) were studied in order to quantify the spinal excitability responsible for spontaneous or induced spasms.

The FR of lower limbs were recorded by surface electrodes applied over the short head of biceps femoris, after electrical stimulation of the ipsilateral sural nerve at the ankle. The stimulating and recording parameters have been described in detail elsewhere [6].

Control responses (in the absence of intrathecal baclofen) were obtained in all 17 patients. Responses during chronic intrathecal baclofen infusion could only be recorded in 15 patients. The following parameters of flexor polysynaptic responses were analysed quantitatively: (a) The *threshold* of FR was defined as the minimal stimulus intensity (in milliampers, mA) necessary to obtain a reproducible FR in the biceps femoris; (b) The *surface* of FR was calculated by integrating the voltage of the electromyographic (EMG) response over an analysis window of 500 msec, the result being expressed in μV x sec. This parameter was calculated for stimulating intensities equal to 2–3 times the reflex threshold; (c) The *duration* of FR (in msec) was estimated between the onset and the end of the reflex response. When the response exceeded the length of the recording window, a duration equal to the window (500 msec) was noted.

VI. Lower Urinary Tract Investigations

Twelve patients with neurogenic bladder dysfunction were evaluated, both in control conditions and under intrathecal baclofen. Standardized clinical assessment of urinary urgency, incontinence (leakage), type of voiding (voluntary, triggered by suprapubic tapping, indwelling or intermittent catheterization) and frequency of urinary infections was done by means of a specially oriented questionnaire. Urodynamic evaluation was performed by cystometry following the technic described before [2]. The bladder was infused with fluid at ambient temperature and recordings of intra-vesical pressure were obtained. During bladder filling, the patient was instructed to report his/her sensations to the examiner. The residual urinary volume was measured after bladder emptying, both immediately before and after micturition. Sphincter EMG recordings were not performed.

Results

a) Effects of Baclofen on Spasticity (Table 2)

Rigidity in the lower limbs was clinically reduced in all patients. The mean *Ashworth* scale score dropped from 3.8 ± 0.7 in the preoperative period to 1.6 ± 0.8 in the last follow-up ($p < 0.001$). The frequency and intensity of the *spasms* were clinically decreased too, the mean spasm score being reduced from 2.8 ± 0.7 preoperatively to 1 ± 0.7 at the last follow-up examination ($p < 0.001$). However, it was frequently observed that spasticity was restored in presence of an urinary infec-

tion or bedsore, or when the drug reservoir level was low ($\leqslant 2$ ml).

b) Effects on Functional Scores, Motor Function and Daily Living Activities (Table 2)

The mean functional score, which initially was 13.4 ± 2.7, decreased after pump implantation to 5.8 ± 2.7 ($p < 0.001$), thus underscoring the overall improvement of the patients for daily-life occupations. *Functional improvement* was dependent on both clinical stage and etiology: it was less marked for tetraparetic and/or multiple sclerosis patients (from 13.9 ± 3.2 to 7.1 ± 1.9) than for paraplegic patients-secondary to spinal cord injury (from 12.5 ± 1.2 to 3.3 ± 2.2). ($p < 0.005$). Previous to pump implantation, fifteen patients were unable to walk, and the sitting position was difficult to maintain, owing to severe rigidity and spasms. In this group, all patients achieved a comfortable sitting position, with improvement of passive mobility and facilitation of transfers, physical therapy and nursing care.

In five patients in whom *motor function* of lower limb was partially preserved previous to treatment, motor improvement occurred as a consequence of the reduction of harmful hypertonia and spasms: one patient among the wheelchair dependent (patient no. 7) became able to walk with crutches after pump implantation; two patients able to stand up and walk a few meters before pump implantation (one with the help of crutches), increased their walking distances after chronic baclofen therapy, but only one of them (patient no. 9) became able to walk alone outside of his house.

Improvement of upper limb motor skills was also observed in four tetraparetic patients, but this was hardly quantifiable with our scales.

c) Effects on Pain

Pain improvement was assessed by the item "pain" of the functional scale (Table 1). Nine patients had *pain related to spasms*, which disappeared after pump implantation in all of them. In 3 cases, spasms disappeared altogether, while in the other 6 patients the frequency and intensity of spasms were significantly decreased and the residual spasms were not painful.

Six patients had *central neurogenic pain*. Three remained unchanged after pump implantation and three seemed improved (reduction of 1 to 2 points in pain scale).

Table 2. *List of Patients Included in the Study*

Case no.	Age/ sex	Diagnosis	Clinical condition	Functional score/10		Spasms score		Ashworth score		Follow-up (month)	Intra-thecal baclof. µg/day
				Initial	Final	Initial	Final	Initial	Final		
1	61 F	SCI	paraplegic	14	3	3	1	2	2	5	132
2	44 F	MS	tetraparetic	15	7	3	2	4	1	28	288
3	39 M	FD	tetraparetic	13	9	3	0	4	1	55	276
4	34 F	SCI	paraplegic	12	1	4	1	4	1	36	204
5	32 F	MS	paraplegic	15	6	2	1	4	1	42	168
6	34 F	FD	tetraparetic	7	3	1	0	3	2	41	168
7	61 M	CM	tetraparetic	10	7	3	2	4	2	24	300
8	25 M	SCI	paraplegic	12	7	4	1	4	1	69	792
9	58 M	SCI	paraparetic	14	2	3	0	5	2	30	365
10	32 F	MS	tetraparetic	14	9	3	1	3	1	6	204
11	52 M	DM	paraplegic	11	5	3	1	4	1	43	184
12	42 M	MS	tetraparetic	16	5	3	0	4	2	47	264
13	39 M	SCI	paraplegic	12	2	3	2	4	2	14	196
14	33 M	MS	tetraparetic	18	9	3	1	5	4	24	384
15	57 M	SY	tetraparetic	12	8	3	1	4	3	55	408
16	52 F	MS	tetraparetic	17	7	2	2	4	1	69	196
17	36 F	MS	paraplegic	16	9	3	2	3	1	45	192
				13,4 ±2,7	5,8* ±2,7	2,8 ±0,7	1* ±0,7	3,8 ±0,7	1,6* ±0,8	36 ±18	277 ±155

SCI spinal cord injury; *MS* multiple sclerosis; *FD* Friedreich disease; *CM* cervical myelopathy; *DM* dorsal myelopathy; *SY* syringomelia; * p < 0,001.

d) *Effects of Baclofen on Lower Limb Flexion Reflexes*

Pre and postoperative flexion reflexes were obtained in 15 out of the 17 patients of the series. Preoperative recordings in the absence of intrathecal baclofen yielded abnormal results in all these patients except two (nos. 7 and 8). Abnormalities always consisted in enhancement of flexion reflexes to exteroceptive stimulation, as shown by significantly higher amplitude and longer duration of responses as compared with normal controls. In about half of the cases (46%) such enhancement was associated with a very significant decrease of the reflex threshold (i.e. reflexes appeared to very small stimulation intensities). In 54% of cases,

flexion reflexes, although significantly enhanced, had a normal or even increased threshold relative to normal controls.

Under intrathecal baclofen, flexion responses were suppressed altogether in the 2 patients whose reflexes were not enhanced preoperatively (nos. 7 and 8). In 3 further cases, reflexes were abolished unilaterally and greatly depressed on the other side (nos. 4, 11 and 17). In the remaining 10 patients, flexion responses remained present albeit significantly decreased in amplitude and duration (Fig. 1). In no patient were responses unchanged under intrathecal baclofen.

No significant correlation appeared between the magnitude of flexor reflexes to sural stimulation and

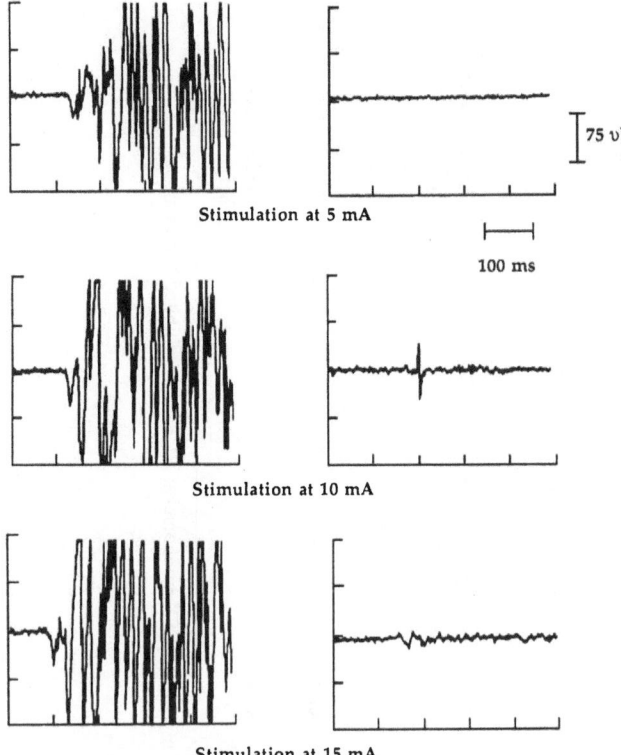

Fig. 1. Abnormal enhanced flexion reflexes to exteroceptive stimulation (5mA), with high amplitude, long duration and appearing at small stimulation intensities. These flexion responses were abolished or strongly attenuated by chronic baclofen infusion, with increasing of their threshold (15mA). Left: control responses, right: under chronic intrathecal baclofen

the number of clinical spasms per unit of time as estimated with the spasm score (see Fig. 2). Clinical spasms reduction and depression of flexion reflexes were not correlated too.

e) Effects on Lower Urinary Tract Function (Table 3)

In three patients, the normal micturition was not modified by intrathecal baclofen. One patient was relieved of his permanent catheter and intermittent self-catheterization became possible, with rare leakage. *Intermittent bladder catheterizations* performed by other five patients were facilitated by intrathecal baclofen owing to reduction of hip adductor hypertonia and spasms.

Urinary urgency in two patients (patients nos. 5 and 16) was remarkably ameliorated under intrathecal baclofen in both cases. *Urine leakage* was present in three cases. After pump implantation, no significant change was noted by one patients, while another was ameliorated and for the third this symptom disappeared (patient no. 16).

The urodynamic quantitative study was performed before and after intrathecal baclofen in 12 of the 14 patients with a neurogenic bladder. In absence of baclofen, nine patients had *hyperactive bladder* characterized by uninhibited contractions with high bladder pressure, elicited at a low volume of urine and associated in some cases with loss of urine. Under intrathecal baclofen, these contractions were suppressed

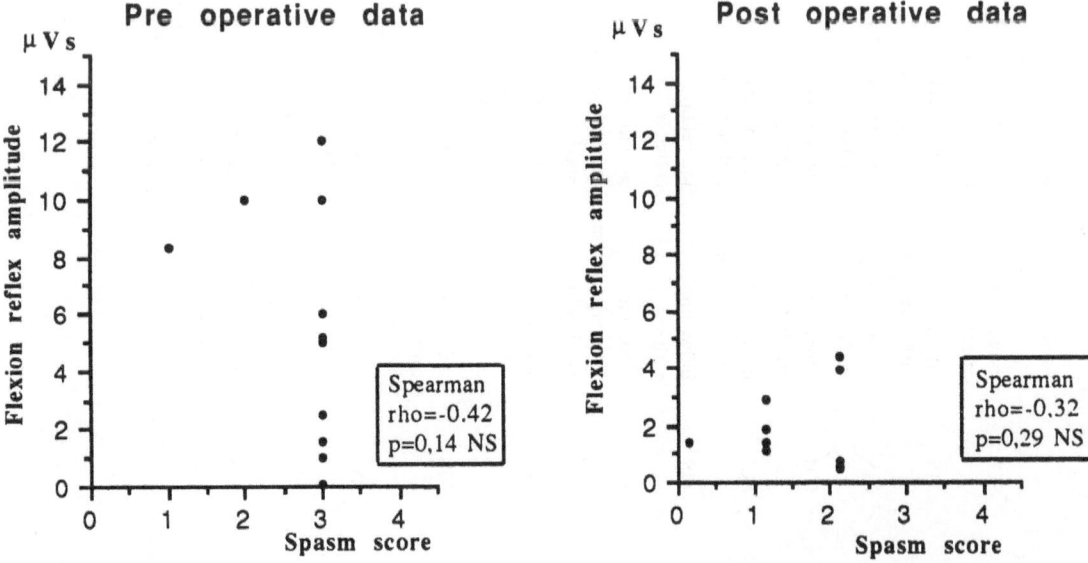

Fig. 2. No significant correlation between exagerate flexion reflex amplitude and spasm scores, before (left) and after (right) chronic intrathecal baclofen. Note that, previous to baclofen, the same clinical score [3] could coexist with a wide range of flexion reflex amplitude. The information provided by flexor reflex was not redundant with that obtained by spasm frequency scale

Table 3. *Effects on Neurogenic Bladder*

Patient no.	Diagnosis	Clinical caracteristcs		Uninhibited contractions				Low compliance bladder		Functional bladder capacity		Remarks
		Before	After	Before Volume ml	Pressure cmH$_2$O	After Volume ml	Pressure cmH$_2$O	Before	After			
1	SCI	indwelling catheter	intermittent catheterisation	141	29	317	18	yes	no	160	470	
2	MS	intermittent catheterisation	intermittent catheterisation	NUC	NUC	NUC	NUC	no	no	500	500	
5	MS	urinary urgency miction by	no symptoms miction by	NUC	NUC	NUC	NUC	no	no	500	500	
7	CM	suprapubic tapping	suprapubic tapping	60	85	116	87	no	no	110	142	
9	SCI	intermittent catheterisation	intermittent catheterisation	NUC	NUC	NUC	NUC	no	no	600	600	
10	MS	indwelling catheter	indwelling catheter	125	54	100	72	yes	yes	125	100	fibrous bladder wall
12	MS	incontinent external collector miction by	incontinent external collector miction by	100	190	100	149	no	no	100	100	external sphincterotomy
13	SCI	suprapubic tapping; leakage	suprapubic tapping; leakage reduction	92	172	195	41	no	no	75	200	
14	MS	indwelling catheter miction by	indwelling catheter miction by	40	44	53	23	yes	yes	50	70	fibrous bladder wall
15	SY	suprapubic tapping	suprapubic tapping	128	116	320	90	yes	no	180	340	
16	MS	urinary urgency; leakage	no symptoms	250	100	NUC	NUC	no	no	250	600	
17	MS	indwelling catheter	indwelling catheter	226	46	NUC	NUC	no	no	226	100	Bladder neck hypotonia

SCI spinal cord injury; *MS* multiple sclerosis; *FD* Friedreich disease; *CM* cervical myelopathy; *DM* dorsal myelopathy; *SY* syringomelia; *NUC* no uninhibited contractions.

in one case (patient no. 16); they occurred at a higher volume in another case (patient no. 7), at a higher volume with lower pressure in three cases (patients nos. 1, 13 and 15) and it were unchanged in three other cases (patients nos. 10, 12 and 14). In case no. 17 a bladder neck hypotonia was verified, with leakage during bladder filling at volume higher to 100 ml. Uninhibited contractions were not observed at this volume.

Four patients had a *low-compliance bladder*, showing abnormal high bladder pressure with step filling curve, related to detrusor hypertonia. In two of them (patients nos. 1 and 15) intrathecal baclofen increased bladder compliance by reducing detrusor hypertonia. In the other two patients (patients nos. 10 and 14), detrusor hypertonia remained unchanged.

Before treatment, nine patients had a reduced *functional bladder capacity* (on average 148 ± 68 ml). Under intrathecal baclofen a significant increase of functional bladder capacity was observed in five patients (402 ± 148 ml, $p = 0.006$). In four others, no significant increase was noticed. It can be explained by a fibrous bladder wall in two cases (patients nos. 10 and 14), by a bladder neck hypotonia in one (patient no. 17) and by an external sphincterotomy in one case (patient no. 12).

f) Complications

Two patients *died* of causes not directly related to baclofen administration. A 61 year-old patient with post-traumatic spinal cord injury died at home of pulmonary embolism, 5 months after pump implantation. In the second case, a 32-year-old woman with advanced multiple sclerosis, the death was caused by aggravation of her underlying disease, 6 months after pump implantation.

Pocket infection was verified in two patients after pump exchange for battery failure. The device was removed and antibiotic therapy introduced with complete recovery. The pump was re-implanted 3 months later in one patient and a neuroablative procedure (lumbo-sacral microsurgical drez-tomy) was performed in the other patient. In this later case the presence of a cystectomy was considered as a additional factor predisposing to infection.

Catheter-related complications were: spontaneous intrathecal catheter rupture (one case), dislogment (one case), radicular pain disappearing after correction of the catheter position (one case) and catheter occlusion (one case). In all these cases, surgical correction was successfully performed under local anaesthesia.

No major *side-effects of intrathecal baclofen* was observed. Mild drug overdose resulting in mental confusion and excessive flaccidity was verified in one patient; it occurred when the patency of system was being tested.

g) Drug Tolerance

Individual dose requirements varied considerably. The increase of dosage to keep the therapeutic effects was evident in the first months after implantation, with relative dose stabilization after the first year. However, in presence of bedsores, urinary infections or multiple sclerosis relapses, the spastic condition was aggravated, prompting a dose increase. Sometimes but not always, this dose could be later reduced. High baclofen doses up to 700–800 µg/day were required by two patients. In the first case, with spinal cord injury, a later stabilization was achieved at 792 µg/day and had lasted for 18 months. In the second case, with cervical syringomyelia, a myelographic study showed a partial blockage of CSF circulation at level of catheter (T12-L1) and the effective dose could be successfully reduced and stabilized at 408 µg/day after catheter replacement.

Discussion

Clinical Findings

Our results confirm the long term efficacy (up to 69 months) of chronic intrathecal baclofen infusion in *reducing muscular hypertonia and spasms*, in agreement with other reports [4,12,18,19,20]. Functional improvement was significantly greater in paraparetic and paraplegic patients with spinal cord injury, than in tetraparetic patients with a progressive disease. The same was also observed by Penn and Kroin 1987 and by Lazorthes *et al.* (1990). In five patients in whom motor function was partially preserved, *motor improvement* was consistently achieved as a consequence of the reduction of harmful hypertonia and spasms. Nevertheless, in the patients able to walk and who were submitted for the test, it was sometimes difficult to reach an optimal equilibrium between reduction of excessive hypertonia and disabling hypotonia. This was the principal cause of exclusion during baclofen trial in our series.

In our series, the *responsiveness of painful spasms* to baclofen was optimal as was the case in previous reports [12,18,21]. *Neurogenic pain was reduced* in three cases and remained unchanged in three other cases. Although the antinociceptive effect of baclofen is supported by experimental studies [22,24], the results in clinical practice are very inconstant: Ochs *et al.* (1989) and Sahuquillo *et al.* (1991) reported that neuropathic pain was unaffected by intrathecal baclofen in their series. On the other hand, Herman *et al.* 1992 reported that intrathecal baclofen suppresses central pain in patients with spinal lesions and Taira *et al.* (1994) in pain after cerebral stroke.

Tolerance

A great variability in the effective dose was observed at 69 months of follow-up. In the present study, no correlation could be found between etiology of spasticity and drug responsiveness or tolerance. Coffey *et al.* 1993 reported that both, the initial dose and the daily dose at last follow-up evaluation, were higher for patients with spinal cord injury than for patients with multiple sclerosis. However, it seems that several factors can be implicated in drug responsiveness: the integrity of nervous structures and the normal cerebrospinal fluid flow [12,18] or the site of drug delivery (position of catheter tip) [8,13]. In addition, technical problems as the occurrence of CSF-leakage at the place where the catheter was introduced in the dural sac has been described simulating drug tolerance [5].

On the other hand, a decrease in the receptor binding density during chronic baclofen infusion was observed in an experimental study [11].

Complications

Catheter-related complications were the most frequently observed complication in our series as in the others [4]. They could be easily corrected under local anaesthesia and without serious consequences.

One paraplegic patient died of pulmonary embolism, in this case the spasms scores was reduced of 3 to 1. It must be questioned for paraplegic patients, that the spasms might not be totally suppressed, to avoid venous stasis and for keeping muscular trophicity in the legs [15,18].

Use of Flexor Recording in the Assessment of Baclofen Efficacy

Clinical spasm scores mainly assess the temporel characteristics of spasms (i.e. recurrence) and their mode of appearence (i.e. spontaneous or induced). In contrast with this, the study of flexion reflexes investigates the intrinsic features of the spasms, namely their *threshold* (i.e. the minimal stimulus intensity required to elicit a spasm) and their *intensity*, as measured by the amplitude and duration of the EMG response in flexor muscles. Both parameters proved to be very significantly influenced by chronic baclofen infusion, which abolished or strongly attenuated flexion responses in all our patients (Fig. 1). In our series, the information provided by flexor reflex recording was not redundant with that obtained by clinical scales: patients with the highest spasm scores were not necessarily those who exhibited the more intense flexor responses to stimulation (see Fig. 2). On the other hand, the intensity of the spasms was better correlated with the patients disability than was by their frequency. Therefore, an independent and specific measure of spasm intensity, such as that provided by flexion reflexes, may prove an useful tool to estimate both the intrusive character of spasms in daily life, and the benefit of antispastic treatment.

Baclofen Effects on Lower Urinary Tract Function

After baclofen infusion, no change was reported by patients with normal miction. In the patients with hyperactive bladder (9 patients), inconstant results was observed. The uninhibited contractions were reduced or suppressed in 50% of cases, even if a optimal control of hypertonia and spasms of lower limbs was present in all patients during urodynamic study. We can suppose that the "effective dose" able to suppress the lower limb spasticity is, in some cases, not enough to achieve an optimal effect on the hyperactive bladder. In experimental studies, the depression of micturition reflex was obtained only with very high doses, probably caused by depression of supraspinal centers (ponto-mesencephalic micturation center) [9,14]. In the other hand, the incomplete suppression of bladder hyperactivity allowed that the suprapubic tapping rest to be effective in voiding bladder (patient nos. 7, 13 and 15).

The increase of functional bladder capacity was observed in only five patients, it can be related to detrusor relaxation and/or to the reduction of the abdominal muscle hypertonia [17]. In the other four patients no improvement was observed: the functional bladder capacity remained restricted due to an inextensible fibrous bladder wall in two cases, and in two other cases due to a permanent urine leakage secondary a bladder neck hypotonia (one case) and to a former external sphincterotomy (one case).

In this series, most of patients assessed were in an advanced stage of disease with serious bladder dysfunction which had not been treated before. It can explain the inconstants results obtained with this treatment in our hands.

Conclusion

The chronic intrathecal baclofen has proved to be an effective and reversible method in the treatment of disabling spasticity of spinal origin, allowing a quantitative modulation of muscular tone of lower limbs. It is especially important for paraparetic patients with useful motricity. For paraplegic patients this method provides a significant amelioration of quality of life secondary to reduction of hypertonia, spasms and pain related to contractures. In the other hand, this treatment is required a rigorous and regular follow-up and a permanent hospital support.

Acknowledgement

This work was supported by the Hospices Civils de Lyon Clinical Research Program, Grant HCL-PMRC no. 94025 on Pain and Spasticity.

References

1. Azouvi P, Roby-Brami A, Biraben A, Thiebaut JB, Thurel C, Bussel B (1993) Effect of intrathecal baclofen on the monosynaptic reflex in humans: evidence for a postsynaptic action. J Neurol Neurosurg Psychiatry 56: 515–519
2. Beneton C, Mertens P, Leriche A, Sindou M (1991) The spastic bladder and its treatment. In: Sindou M, Abbott R, Keravel Y (eds) Neurosurgery for spasticity. Springer, Wien New York, pp 193–199
3. Bowery NG, Hill DR, Hudson AL (1983) Characteristics of GABA B receptor binding sites on rat whole brain synaptic membranes. Brit J Pharmacol 78: 131–120
4. Coffey RJ, Cahill D, Steers W, Park TS, Ordia J, et al (1993) Intrathecal baclofen for intractable spasticity of spinal origin: results of a long-term multicenter study. J Neurosurg 78: 226–232
5. Delhaas EM, Brouwers JRBJ (1991) Intrathecal baclofen overdose: report of 7 events in 5 patients and review of the literature. Int J Clin Pharm Th 29: 274–328
6. Garcia-Larrea L, Sindou M, Mauguière F (1989) Nociceptive flexion reflexes during analgesic neurostimulation in man. Pain 39: 145–156
7. Herman RM, Luzansky SCD, Ippolito R (1992) Intrathecal baclofen suppresses central pain in patients with spinal lesions. Clin J Pain 8: 338–345
8. Hugenholtz H, Nelson RF, Deoux E (1993) Intrathecal baclofen-the importance of catheter position. Can J Neurol Sci 20: 165–167
9. Kums JJM, Delhaas EM (1991) Intrathecal baclofen infusion in patients with spasticity and neurogenic bladder disease. World J Urol 9: 99–104
10. Kroin JS, Penn RD, Beissinger RL, Arzbaecher RC (1984) Reduced spinal reflexes following intrathecal baclofen in the rabbit. Exp Brain Res 54: 191–194
11. Kroin JS, Bianchi GD, Penn RD (1993) Intrathecal baclofen down-regulates GABA-B receptors in the rat substantia gelatinosa. J Neurosurg 79: 544–549
12. Lazorthes Y, Sallerin-Caute B, Verdie JC, Bastide R, Carillo JP, (1990) Chronic intrathecal baclofen administration for control of severe spasticity. J. Neurosurg 72: 393–402
13. Loubser PG, Narayan RK (1993) Effect of subarachnoid catheter position on the efficacy of intrathecal baclofen for spasticity. Anesthesiology 79: 611–614
14. Margora F, Shazar N, Drenger B (1989) Urodynamic studies after intrathecal administration of baclofen and morphine in dogs. J Urol 141: 143–147
15. Mertens P, Millet MF, Sindou M (1990) Traitement de la spasticité par infusion intrathecale de baclofen au moyen de pompes implantables. In: Pelisser J, Herison C (eds) Actual Med Phys Reeduc 15e serie Simon. Masson, Paris, pp 386–392
16. Müller H, Zierski J, Dralle D, Börner U, Hoffmann O (1987) The effect of intrathecal baclofen on electrical muscle activity in spasticity. J Neurol 234: 348–352
17. Nanninga J, Frost F, Penn R (1989) Effect of intrathecal baclofen on bladder and sphincter function. J Urolol 142: 101–105
18. Ochs G, Struppler A, Meyerson BA, Linderoth B, Gybels J, et al (1989) Intrathecal baclofen for long-term treatment of spasticity: a multi-center study. J Neurol Neurosurg Psychiatry 52: 933–939
19. Penn RD, Savoy SM, Corcos D, Latash M, Gottlieb G, et al (1989) Intrathecal baclofen for severe spinal spasticity. N Engl J Med 230: 1517–1521
20. Penn RD, Kroin JS (1987) Long-term intrathecal baclofen infusion for treatment of spasticity. J Neurosurg 66: 181–185
21. Sahuquillo J, Muxi T, Noguer M, Jodar R, Closa C, et al (1991) Intraspinal baclofen in the treatment of severe spasticity and spasms. Acta Neurochir (Wien) 110: 166–173
22. Sabbe MB, Grafe MR, Pfeifer BL, Mirzai THM, Yaksh T (1993) Toxicology of baclofen continuously infused into the spinal intrathecal space of the dog. Neurotoxicology 14: 397–410
23. Taira T, Tanikawa T, Kawamura H, Iseki H, Takakura K (1994) Spinal intrathecal baclofen suppresses central pain after stroke. J Neurol Neurosurg Psychiatry 57: 381–386
24. Wilson PR, Yaksh TL (1978) Baclofen is antinociceptive in the intrathecal space of animals. Eur J Pharmacol 51: 323–330

Correspondence: Patrick Mertens, M.D., Neurochirurgie A, Hôpital Neurologique, 59 Boulevard Pinel, 69003 Lyon, France.

Acta Neurochir (1995) [Suppl] 64: 26–29

A Neurophysiological Method for the Evaluation of Motor Performance in Spastic Walking Patients

I. Dones, D. Servello, F. Molteni[1], G. Mariani[1], and G. Broggi

Department of Neurosurgery, Istituto Nazionale Neurologico "C. Besta", Milano, Italy, and [1]Department of Motor Rehabilitation, Ospedale Valduce, Como, Italy

Summary

Intrathecal baclofen is at present the best treatment for severe spasticity of various etiologies. In walking patients affected by severe spasticity a careful evaluation of the motor performance is needed for a correct indication for this treatment. The examination should focus on the delicate balance between spasticity and voluntary muscle activation which is crucial for an improvement of motor performance during gait. Seven patients have been neurophysiologically evaluated by the use of a Cibex apparatus measuring torque and movement velocity of the lower limbs simultaneously with static and dynamic recordings of the EMG.

Keywords: Spasticity; intrathecal baclofen.

Introduction

Spasticity is a clinical sign often present in association with decreased muscle strength. It is characterized by a marked increase in muscle tone both of agonist and antagonist muscles typically giving progressive resistance to a fast passive movement of the limb. In patients who are bedridden or confined to a wheelchair spasticity resulting in exaggerated extension of the legs severely interferes with nursing and care, whereas in patients who have some retained walking capacity spasticity requires the usage of crutches making the walking unstable and hinders the climbing of stairs.

Severe spasticity can be effectively treated with intrathecal baclofen administered by means of a sophisticated but expensive programmable and implantable pump. The implantation of such a device is performed only when there is no response to any oral antispastic treatment without side effects and there is a positive response to test administration of bolus doses of intrathecal baclofen.

This is a report of the usage of a neurophysiological test battery for the evaluation of motor performance in spastic walking patients who may be candidates for permanent intrathecal baclofen treatment. The clinical evaluation was based on the Ashworth scale of spasticity and the scales of reflexes and of muscle spasms (Table 1) which are in use in the international protocol for the treatment of spastic patients with intrathecal baclofen.

Patients, Material and Methods

Seven patients ranging from 24 to 50 years of age and with a history of familial spastic paraparesis ranging from 2 to 20 year were included in the study: one of the patients was diagnosed as having juvenile amyotrophic lateral sclerosis. The spasticity in these patients ranged from grade 3 to grade 5 in the Ashworth scale and they were all able to walk with the help of two crutches but they were unable to climb stairs. All patients had tried antispastic treatment with oral baclofen or tizanidine without serious side-effects but with insufficient benefit (Table 2).

The patients were subjected to a battery of neurophysiological tests before and following a bolus dose of intrathecal baclofen.

In order to assess lower limb spasticity, the torque and angular range of motion (ROM) were measured together with the recording of dynamic EMG from quadriceps and hamstrings muscles during alternate flexion-extension movements of the knee. A Cybex 6000 Isokinetic Dynamometer (Technogym, Italy) was used for the ROM measurements. Surface EMG electrodes were used and the signals were filtered, rectified and processed using finite impulse response (FIR) filters. The patients were comfortably seated in the Cybex chair with the backrest set at 90°. Trunk and thigh were firmly fixed with straps.

For the examination of the continue passive motion (CPM) the patient was asked to relax as much as he could while the dynamometer lever moved passively the lower limb from 90° of knee flexion to complete knee extension at constant velocity. The axis of rotation of the dynamometer was aligned to the axis of rotation of the knee joint. Three sessions of three movement cycles each with angular velocities of 60°/sec, and 180°/sec. were performed. The

Table 1. *The Ashworth Scale of Spasticity, the Spasm Frequency Scale and the Scale for Tendon Reflexes Used for the Clinical Evaluation and Scoring of Spastic Patients*

Grade	Muscle tone
Ashworth scale for spasticity	
1	normal
2	slight increase in tone
3	more marked tone, but affected parts easily flexed
4	considerable increase in muscle tone, difficulties in passive movements
5	affected parts rigid in flexion or extension
Spasm frequency scale	
0	none
1	no spontaneous spasms, vigorous sensory and motor stimulation results in spasms
2	occasional spontaneous spasms
3	1–10 spontaneous spasms per hour
4	more than 10 spontaneous spasms per hour
Scale of "tendon" reflexes	
0	absent
1	normal
2	increased
3	clonus

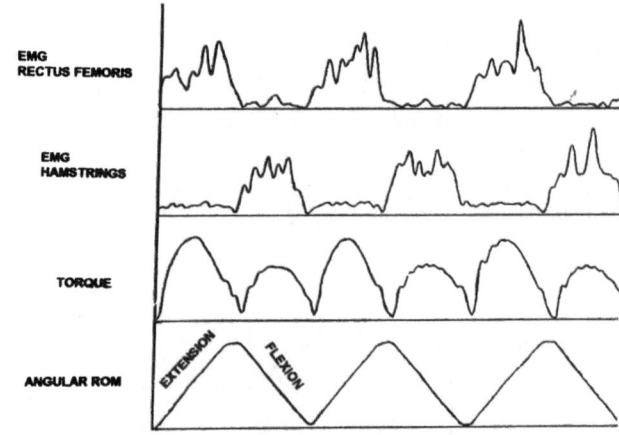

Fig. 1. Records of movements and muscle activation in a normal subject

(3) the EMG activity of the rectus femoris and hamstring muscles were filtered and rectified and analyzed in relation to the recorded torques.

Results

In a control, non spastic subject, the pattern of motion and muscle activation during active movement at 120°/sec angular velocity is shown in Fig. 1.

In spastic patients three distinctly different types of disturbances could be distinguished:

First type: decreased stretch refex activation during the lengthening of the quadriceps and hamstring muscles.

Second type: deficient voluntary muscle activation with decreased or absent motor unit recruitment.

Third type: the performance of active movements is mainly hampered by the co-activation of agonistic and antagonistic muscles.

Fourth type: mixture of the above mentioned disturbances.

The different types of spasticity-related disturbances are exemplified by the following three patients.

recorded torques and the EMG activity produced during muscular lengthening allowed the assessment of the velocity-dependent stretch reflexes of the flexor and extensor muscles.

In second phase of the procedure three sessions of three repetitive voluntary knee flexion-extension movements with maximal effort were performed. Since the patients were unable to produce consistent torques to pull the lever through the complete ROM in a isokinetic way, we set the Cybex dynamometer with an active-assisted program. In this way, the lever, moving at a preset constant velocity, helped the leg to perform the full range of motion. Different angular velocities – 60°/sec., 120°/sec. and 180°/sec – were applied.

In the dynamic phase of the test the following data were collected:
(1) the maximal extensor and flexor torques produced at different velocities. The recorded torque is the resultant of two movement forces: one is due to the activation of agonist muscles and the other results from contraction of antagonistic muscles acting in the opposite direction. Articular stiffness adds to these forces. Torque-velocity linear regression ratio, which in normal subjects has a decreasing pattern, was analyzed;
(2) the mean power and the power-velocity ratio were analyzed;

Table 2. *Patients Selected for the Neurophysiological Evaluation Before and After Intrathecal Baclofen.* Walking spastic patients selected for passive and active neurophysiological assessment before and after intrathecal baclofen

Pt.	Age	Disease	Ashworth	Condition
1	50	FSP	5	ambulation with two canes by swinging movements
2	34	FSP	5	poor walking with two canes
3	50	FSP	3	walking but not climbing stairs
4	32	FSP	5	moving with great difficulty with two canes
5	24	ALS	4	walking with one cane, not climbing stairs
6	45	FSP	4	walking with difficulty on flat surfaces
7	47	FSP	5	walking with two canes

FPS familial spastic paraplegia; *ALS* amyotrophic lateral sclerosis. The condition and the Ashworth scoring are measured before treatment.

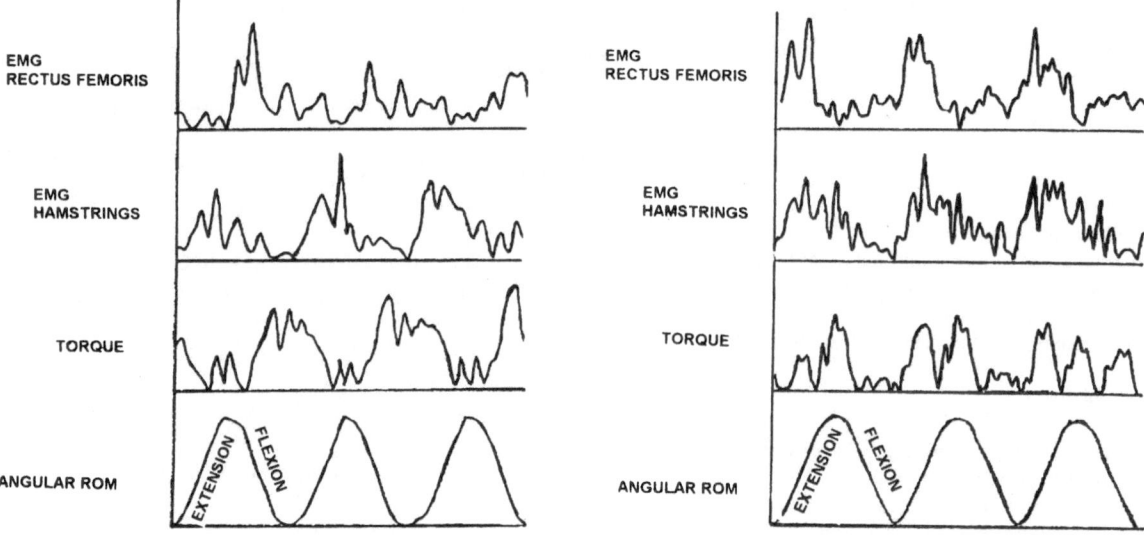

Fig. 2. Case 1: Records of movement and muscular activation before and after a bolus test with intrathecal baclofen

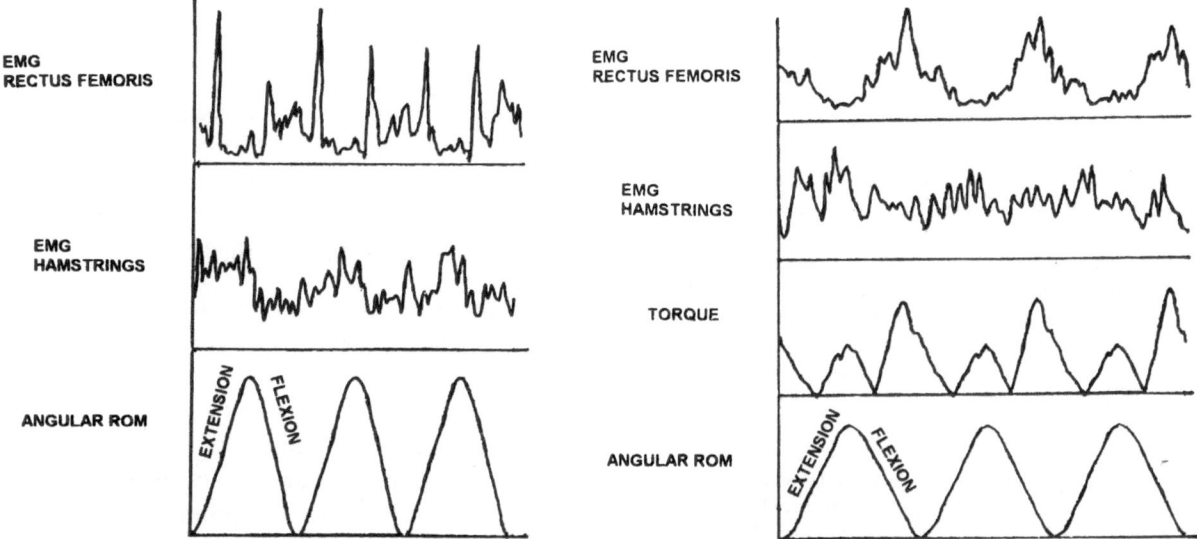

Fig. 3. Case 2: Records of movements and muscular activation during passive and active movements

Case 1. Male, familial spastic paraparesis, 50 years old, 25 years of disease history, grade 5 of the Ashworth scale. By analizing the biomechanical and electromyographic data the spastic paraparesis this case can be characterized as the first type where the main feature was the low stretch reflex thresholds. A bolus administration of intrathecal baclofen (25 µg) produced a significant bilateral decrease ·of stretch reflex activation particularly in the extensor muscles. There was an improved pattern of recruitment of knee extensor units during voluntary contractions at low velocity (60°/sec) but at high velocity (180°/sec) the interference of the enhanced stretch reflexes of the extensor muscles was

unchanged. The patient was later implanted with a pump and is now able, with 75 micrograms/day of intrathecal baclofen, to walk with only one cane and to climb stairs without assistance (Fig. 2).

Case 2. Female, familial spastic paraparesis, 34 years old, 16 years history of the disease, grade 5 of the Ashworth scale. The knee flexion-extension movement was highly impaired by stiffness particularly in the extensor muscles. The spasticity had a mixed pattern (type 4) with both a weak stretch reflex and a lack of voluntary muscular activation (Fig. 3).

This patient had previously had a pump implanted on the basis of a general clinical examination. However,

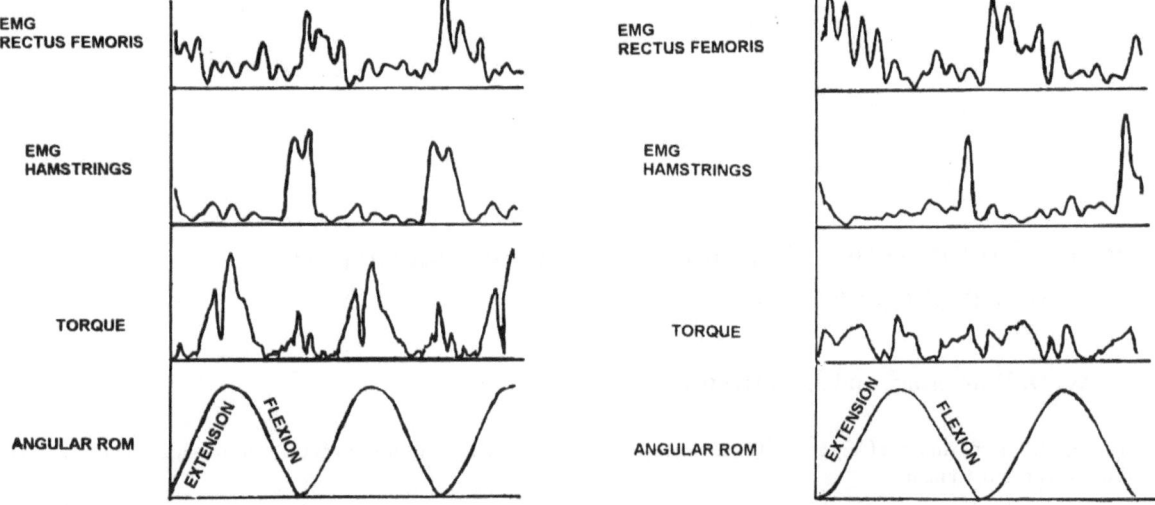

Fig. 4. Case 3: Records of movements and muscular activation during passive and active movement

the administration of a test dose of 25 micrograms/day revealed that the drug caused an impairment of the motor function. The treatment was therefore withdrawn.

Case 3. Male, Familial spastic paraparesis, 45 years old, 25 years history of the disease, grade 4 of the Ashworth scale. In this case, the pattern of spasticity was classified as the first type referred to above. The EMG traces of the CPM phase showed a moderate intensity of the stretch reflex both in the rectus femoris and in the hamstrings muscles. The EMG traces of the dynamic phase showed a good timing of flexor-extensor muscle activation with a low level of co-contraction (Fig. 4).

A bolus test dose of 75 micrograms of baclofen resulted in an improved motor performance and the patient was considered a good candidate for continued baclofen treatment via intrathecal administration.

Discussion

In patients who do not benefit from oral baclofen treatment and who are bedridden or confined to a wheelchair, the decision to perform an implantation of a pump for continuous intrathecal administration may be based on a clinical evaluation of the spasticity using the scales referred to above and the response to a bolus test dose of baclofen. In these patients the prime goal is to reduce their spasticity and muscle spasms and the possible effect on residual voluntary motor performance can generally be disregarded.

In spastic patients who are still able to walk the prime aim of the treatment is to improve motor performance. In our experience, the clinical effect of test doses of intrathecal baclofen is difficult to evaluate and do not provide reliable guidance for the decision of implanting an expensive device for continued treatment. For such patients, we believe that the neurophysiological evaluation as reported in the study may be a valuable tool in determining whether a patient can be expected to enjoy a substantial improvement of the motor performance by the reduction of the spasticity. Moreover, we have had the experience that our neurophysiological tests can be useful in guidening the daily dosage of intrathecal baclofen. The prime advantage with our tests is not to measure spasticity per se, but to give a ratio between spasticity and muscle strength during both passive and active movements of the limb providing reliable information about the functioning of the limbs during walking.

Correspondence: Giovanni Broggi, M.D., Department of Neurosurgery, Institute Nazionale Neurologico "C. Besta", Via Celoria, 11, VI-20133 Milano, Italy.

Acta Neurochir (1995) [Suppl] 64: 30–34

Microelectrode Monitoring of Cortical and Subcortical Structures During Stereotactic Surgery

A. M. Lozano[1], **W. D. Hutchison**[2], and **J. O. Dostrovsky**[2]

Division of Neurosurgery, [1]Department of Surgery, [2]Department of Physiology, The Toronto Hospital Neurological Centre and The University of Toronto, Toronto, Canada

Summary

We describe microelectrode recording and stimulation techniques to delineate the cellular boundaries and neural projections of stereotactic brain targets. These techniques have applications in the surgery for pain, movement disorders and in psychosurgery. Neuronal records from stereotactic operations including thalamotomy, pallidotomy, cingulotomy and anterior capsulotomy are discussed. These tools are used to distinguish gray matter from white matter, to obtain direct measures of cellular activity in the target, to identify the physiological properties and receptive fields of the subpopulation of neurons at the electrode tip and to avoid lesion making induced injury to adjacent structures. Microelectrode recording and stimulation techniques improve physiological localization and decrease the possibility of unwanted neurological complications with functional stereotactic procedures.

Keywords: Globus pallidus; psychosurgery; pain; Parkinson's disease.

Introduction

The accurate identification of subcortical and cortical structures during stereotactic surgery is essential in guiding the placement of lesions for therapeutic benefit. The determination of targets by imaging techniques alone is subject to variable accuracy and inter-individual differences in anatomy and physiology. Because of this, the confirmation of stereotactic targets using neurophysiological landmarks is important to maximize the benefits and minimize unwanted neurological complications.

We employ microelectrode recording and stimulation techniques in all functional neurosurgical operations which at our centre include pallidotomy for movement disorders, mesencephalic tractotomy, thalamotomy and deep brain stimulation, for pain and movement disorders and cingulotomy and anterior capsulotomy for obsessive compulsive disorders and other psychiatric disturbances and chronic intractable pain.

Each procedure has in common the necessity to distinguish gray and white matter and to define the nature of the surgical target and the important adjacent structures. The microelectrode recording techniques make it possible to record from individual neurons and test their responsiveness to various somatosensory stimuli. In addition, the passage of small electrical currents through the fine tip of the microelectrodes allows spatially precise localization of sensations, movements and evoked potentials. This permits a physiological definition of the target and identifies important neighbouring structures to be protected.

This paper briefly describes the microelectrode technique and illustrates some typical properties and characteristic responses of selected neurons.

Methods and Materials

Stereotactic Procedure

All operations are conducted under local anaesthesia. A Leksell model G stereotactic frame is used. Patients undergo either CT or MRI in stereotactic conditions to calculate the coordinates of the target directly or in relation to the position of the anterior and posterior commissures. Where applicable, digitized sagittal brain maps from a stereotactic atlas [13] are stretched or shrunken according to the relation between the patient's intercommissural line and that of the atlas.

Microelectrode Recording

Laboratory assembled microelectrodes including commercially available parylene-C insulated tungsten microelectrodes are adapted for extracellular recording as previously discribed [3,7,17]. The exposed tips of the electrodes are electroplated with gold and platinum to reduce the impedance to 0.5 to 1 MegaOhm to improve the signal to noise ratio of recordings.

The microelectrodes are inserted into protective 19 gauge stainless steel guide tubes that are held on the arc of the stereotactic frame. The electrode is extruded into the brain for up to 20 mm beyond the end of the guide tube in small increments using a hydraulic microdrive.

Single unit activity detected by the microelectrode is amplified (DAM 80, World Precision Instruments) and filtered (200–5000 K) to remove unwanted frequencies. Tracings are displayed on an oscilloscope and fed to an audio monitor. Data are stored on videotape using an 8 channel digital recording device (Instrutech).

Microelectrode Stimulation

The delivery of small electric currents through the microelectrode tip is carried out with a constant current stimulus isolation unit and pulse generator (World Precision Instruments). The stimulation parameters normally used are 1 sec trains at 300 Hz with a 0.2 ms pulse width and currents from 1 to 100 microamps. In some instances longer trains of up to 4 sec are used.

Visual Evoked Potentials

The optic tract is identified by responses to microstimulation and by recording axonal multiunit activity and/or slow wave evoked potentials to repetitive flashes from a strobe light.

Results and Discussion

Ventral Intermediate Nucleus (VIM) of the Thalamus

Lesions or chronic electrical stimulation of VIM thalamus are used to treat tremor associated with Parkinson's disease, essential tremor, cerebellar tremors or post traumatic tremors [3,10–12,17]. VIM thalamotomy has also been used in dystonia [19]. VIM targets are generally 14–17 mm from the midline, 2 to 8 mm anterior to the posterior commissure and 0 to 3 mm above the intercommisural line (Fig. 1A).

As studied by microelectrode recordings, VIM neurons produce large amplitude potentials and have a high spontaneous activity. VIM neurons respond primarily to passive (and voluntary) movements. In addition, these neurons respond to tapping and squeezing of the muscles or tendons. VIM neuronal responses may be phasic or tonic. This nucleus is arranged somatotopically with neurons subserving intraoral structures positioned medially and the upper limb and lower limb progressively more laterally.

A characteristic of VIM is the presence of neurons that fire in bursts that are synchronous with peripheral tremor (Fig. 2). Microstimulation at the site of these "tremor cells" can produce an almost instantaneous arrest in tremor that usually returns soon after the period of stimulation has ceased.

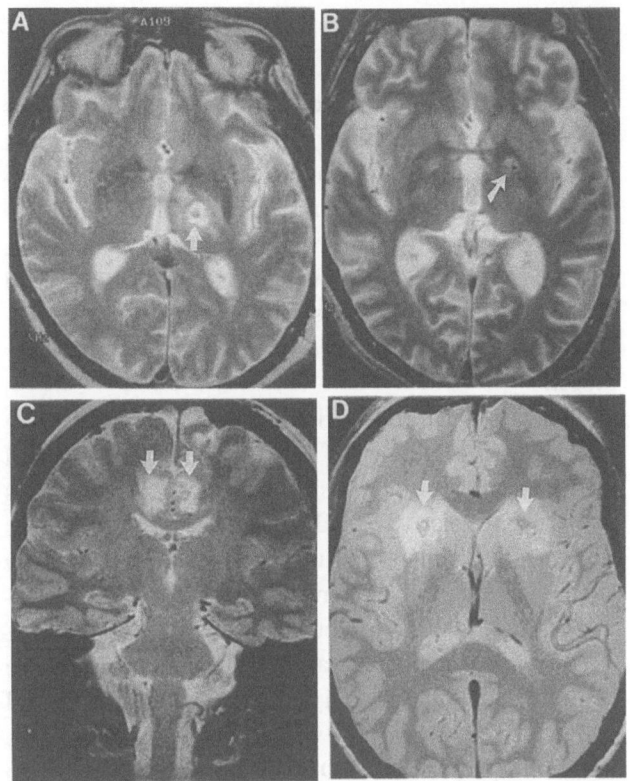

Fig. 1. Postoperative MRI scans showing (A) thalamotomy for tremor, (B) posteroventral GPi pallidotomy, (C) cingulotomy and (D) anterior capsulotomy. Lesions and surrounding edema are indicated by white arrows. The MRIs in (A,C) and (D) were done at 0–7 postoperative days; (B) at 3 months

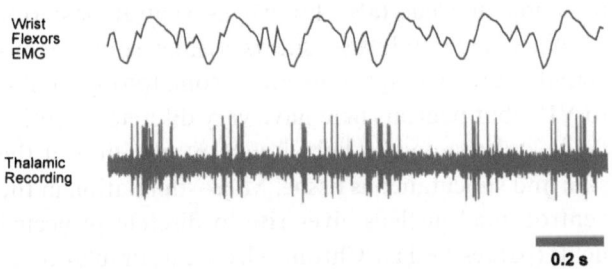

Fig. 2. Thalamic tremor cell. Recording of a neuron in VIM in a Parkinson's patient. This example shows a neuron in VIM thalamus whose firing is time locked with wrist tremor as shown in the EMG tracing

There is increasing interest in the use of chronic electrical stimulation of VIM to treat tremor because this modality is showing promising effectiveness and may be associated with a lower risk than lesion surgery [1]. The surgical target for chronic stimulation appears to be the same as that for lesioning.

The criteria for target selection in VIM for the placement of chronic stimulating electrodes or lesions include: an area populated by neurons rather than axons, an area anterior to the ventrocaudal nucleus and superior to the base of the thalamus, an area populated by neurons which respond to joint movements (Kinesthetic cells) and which discharge at a frequency similar to peripheral tremor (tremor cells, Fig. 2), an area where the mechanical disturbance with the microelectrode diminishes tremor and where electrical microstimulation produces tremor reduction or arrest.

There are a number of risks associated with creating lesions in the vicinity of VIM. Paresis occurs with injury to the laterally positioned internal capsule. As studied with microelectrodes, the internal capsule does not contain high amplitude units and is relatively silent. Microstimulation in the capsule produces paraesthesia or motor contractions.

Ataxia and hypotonia after VIM thalamotomy are thought to be related to dysfunction of the ventrooral posterior (VOP) nucleus which receives the dentatothalamic projection. VOP is situated anterior to VIM and contains "voluntary neurons" which fire in advance of limb movements and are believed to be involved in aspects of the planning of movements.

Numbness occurs if the VIM lesion encroaches on the somatosensory ventrocaudal nucleus (immediately posterior to VIM) or its afferent fibres. The ventrocaudal nucleus (also known as ventral posterior medial and lateral) also shows a high spontaneous activity, large voltage units and a somatotopy similar to VIM but neurons here have very discrete receptive fields and respond to light touch or vibration of the skin and subcutaneous tissue. Microstimulation in the ventrocaudal nucleus gives rise to discrete projected fields (paraesthesias). Chronic electrical stimulation of this sensory nucleus is sometimes used to treat chronic pain particularly central and deafferentation pain.

Globus Pallidus

Lesions in the globus pallidus (Fig. 1B) are used in the treatment of bradykinesia and rigidity in Parkinson's disease [9,16]. The surgical procedure is believed to work by downregulating the tonic inhibitory overactivity of the internal segment of the globus pallidus (Gpi) on the motor system [4,5,18]. The published target is 18–22 mm lateral to the midline, 3 mm anterior to the midcommissural line and 6 mm below the intercommissural line [9]. A recent report suggests that chronic electrical stimulation of the internal segment of the globus pallidus may have similar benefits [14].

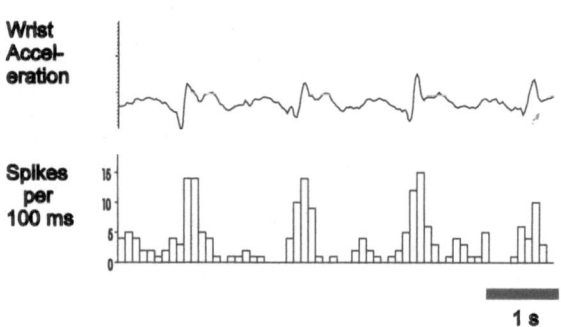

Fig. 3. Movement-related unit in globus pallidus. Pallidal neurons respond to limb movements. Firing rate histogram of a single GPi neuron whose discharge pattern is related to wrist movement as indicated by the wrist accelerometer

For pallidotomy, the trajectory of the microelectrode is 19 to 21 mm from the midline. The microelectrode passes successively through the putamen, the external globus pallidus (Gpe) the external and internal segments of the internal globus pallidus (GPie, Gpii) and the optic tract. GPie and GPii are separated by a 1 mm wide lamina of white matter which can often be recognized as an area of diminished activity.

In patients with Parkinson's disease, GPe neurons show 2 distinct discharge patterns [7,15]. Some units have a slow frequency discharge (10 to 20 Hz) interrupted by pauses, others discharge at a higher frequency (30 to 60 Hz) also with intervening pauses.

Neurons in GPi have a baseline rate of firing that is higher than that found in GPe [7,15]. The range of discharge rates is 20 to 200 Hz with an average rate of approximately 80 Hz and with few prolonged pause periods.

Both GPi (Fig. 3) and GPe units respond to joint movements. Although the majority of units respond exclusively to contralateral movements, several units can be activated by movements of both right and left limbs. Units can show a preference to voluntary or passive movements.

Border cells are a third class of neurons in the globus pallidus. In primates, these neurons are situated at the borders of the internal and external pallidal segments [2]. Border cells fire with a regular pattern with a rate of 30 to 40 Hz [7]. In contrast to neurons in GPi and GPe, we have not found that the discharge pattern of border neurons is influenced by movements.

It is important to identify the optic tract during pallidotomy to identify the ventral border to the target and to avoid possible visual complications of lesion making. The microelectrode entering the optic tract is signalled by a loss of detected cellular activity, by the activation of axonal discharges and visual evoked potentials in a time-related fashion to strobe light stimulation and by the observation of electrical stimulation effects. Microelectrode stimulation is done using 200 microsecond pulses at 1 to 100 microamps and 300 Hz. Patients report blue, yellow or white lights; points, stars or clouds in the contralateral visual field. The threshold for stimulation can be as low as 2 microamps. Increasing the stimulating current produces a larger and brighter visual perception. With increasing distance away from the optic tract, the stimulation thresholds increase. We do not use currents above 100 microamps. The distance between the most ventral GPi unit and the optic response is generally 1 to 2 mm.

The criteria for making lesions in the internal globus pallidus of patients with Parkinson's disease are: 1) the identification of units in GPi that responded to movements to confirm target localization and 2) the identification of the optic tract and internal capsule to delineate the ventral and posterior borders of GPi and to ensure that these structures can be spared.

The customary GPi lesion we make is approximately 6 mm in diameter and is placed in an area of GPi that contains neurons responding to joint movements that are at least 4 mm dorsal to the optic tract.

Cingulate Gyrus

Cingulotomy is performed for obsessive compulsive syndrome and in certain chronic pain states [6,8]. The published targets are in the cingulate gyrus from 2 to 4 cm behind the anterior border of the lateral ventricles. The cingulum can be targeted directly from the MR image (Fig. 1C).

Microelectrode recordings in the cingulum reveal neuronal activity in the dorsal portion of the cingulate gyrus. As the electrode courses ventrally, the cingulate bundle shows low background noise which is characteristic of white matter. The cortical layer at the bottom of the cingulate gyrus is signalled by a reappearance of neuronal action potentials. Finally, the neural activity returns to a quiet background as the electrode leaves the cingulate gyrus to penetrate the underlying corpus callosum.

We have not noted any effects of electrical stimulation in the cingulum with low currents.

Anterior Limb of the Internal Capsule

Anterior capsulotomy is used in the treatment of psychiatric disorders including obsessive compulsive

disorder, affective disorders and intractable pain. The recommended targets are 17 mm rostral to the anterior commissure, 18–20 mm lateral to the midline and based from the level of the intercommissural line to a point 16 mm in a rostral and dorsolateral direction [8]. The capsulotomy target is easily selected from the MR image (Fig. 1D).

Microelectrode recordings during capsulotomy reveal the presence of spontaneously active low firing frequency units in the head of the caudate nucleus lying immediately medial to the capsule at the lesion base. Within the capsule itself there is low background noise and no neuronal action potentials. We have not noted any effects of electrical stimulation of the anterior limb of the internal capsule using currents up to 100 microamps.

Conclusion

Modern stereotactic imaging and neurophysiological techniques makes possible the precise physiological definition of stereotactic brain targets. These studies provide important information on neuronal function and increase the accuracy and safety of stereotactic procedures. Further refinements of these techniques will define the optimal physiologic targets for stereotactic neurosurgery. With chronic stimulation procedures, the end-point will be to choose the most effective target where stimulation is free of unwanted side effects. In lesion surgery, the goal will be to define the smallest possible lesion that is therapeutically effective and completely devoid of unwanted complications.

References

1. Benabid AL, Pollak P, Gervason C, et al (1991) Long-term suppression of tremor by chronic stimulation of the ventral intermediate thalamic nucleus. Lancet 337: 403–406
2. Delong M (1971) Activity of pallidal neurons during movement. J Neurophysiol 34: 414–427
3. Dostrovsky JO, Sher GD, Davis KD, Parrent AG, Hutchison WD (1993) Microinjections of lidocaine in the human thalamus: a useful tool in stereotactic surgery. Stereotact Funct Neurosurg 60: 168–174
4. Filion M, Tremblay L, Bedard PJ (1988) Abnormal influences of passive limb movement on the activity of globus pallidus neurons in parkinsonian monkeys. Brain Research 444: 165–176
5. Filion M, Tremblay L (1991) Abnormal spontaneous activity of globus pallidus neurons in monkeys with MPTP-induced parkinsonism. Brain Res 547: 142–151
6. Hassenbusch SJ, Pillay P, Barnett GH (1990) Radiofrequency Cingulotomy for intractable cancer pain using stereotaxis guided by magnetic resonance imaging. Neurosurgery 27: 220–223
7. Hutchison W, Lozano AM, Davis K, et al (1994) Differential

neuronal activity in segments of globus pallidus in Parkinson's disease patients. Neuroreport 5: 1533–1537

8. Laitinen LV (1988) Psychosurgery today. Acta Neurochir (Wien) [Suppl] 44: 158–162
9. Laitinen LV, Bergenheim AT, Hariz MT (1992) Leksell's posteroventral pallidotomy in the treatment of Parkinson's disease. J Neurosurg 76: 53–61
10. Narabayashi H (1989) Stereotaxic VIM thalamotomy for treatment of tremor. Review Eur Neurol 29 [Suppl 1]: 29–32
11. Ohye C, Shibazaki T, Hirai T, *et al* (1989) Further physiological observations on the ventralis intermedius neurons in the human thalamus. J Neurophysiol 61: 488–500
12. Ohye C, Shibazaki T, Hirato M, *et al* (1990) Strategy of selective VIM thalamotomy guided by microrecording. Stereotact Funct Neurosurg 54–55: 186–191
13. Schaltenbrand G, Bailey P (1959) Einführung in die Stereotaktischen Operationen mit einem Atlas des menschlichen Gehirns, Vol 3. Thieme, Stuttgart
14. Siegfried J, Lippitz B (1994) Bilateral chronic electrostimulation of the ventroposterolateral pallidum: a new therapeutic approach for alleviating all parkinsonian symptoms. Neurosurgery 35: 1126–1130
15. Sterio D, Beric A, Dogali M, *et al* (1994) Neurophysiological properties of pallidal neurons in Parkinson's disease. Ann Neurol 35: 586–591
16. Svennilison E, Torvik A, Lowe R, Leksell L (1960) Treatment of parkinsonism by stereotactic thermolesions in the pallidal region. Acta Psychiat Scand 35: 358–377
17. Tasker RR, Organ LW, Harylshyn PA (1982) The thalamus and midbrain of man: a physiological atlas using electrical stimulation. Thomas, Springfield, Ill
18. Wichmann T, Delong MR (1993) Pathophysiology of parkinsonian motor abnormalities. Adv Neurol 60: 53–61
19. Yamashiro K, Tasker RR (1993) Stereotactic thalamotomy for dystonic patients. Stereotact Funct Neurosurg 60: 81–85

Correspondence: Andres M. Lozano, MD, PhD, FRCS(C), Division of Neurosurgery, 2–433 McLaughlin Pavilion, Toronto Western Hospital, 399 Bathurst St. Toronto, M5T 2S8, Canada.

Acta Neurochir (1995) [Suppl] 64: 35–39

Neurophysiological Monitoring of Cranial Nerves During Posterior Fossa Surgery

G. Broggi[1], **V. Scaioli**[2], **S. Brock**[1], and **I. Dones**[1]

Departments of [1]Neurosurgery and [2]Neurophysiology, Istituto Nazionale Neurologico "C. Besta", Milano Italy

Summary

Intraoperative neurophysiological monitoring of cranial nerve functions in surgery for microvascular decompression and tumors of the posterior fossa is important for minimizing risk of permanent damage to the nerves. In particular, intraoperative BAEP and the EMG function of muscle innervated by trigeminal and facial muscle have been found useful. We report here our experiences with intraoperative monitoring of brainstem auditory evoked potentials (BAEP) and EMG recorded from muscles supplied by the trigeminal and facial nerves.

Keywords: Cranial nerves; posterior fossa surgery.

Introduction

Patients undergoing posterior fossa surgery for tumor removal or microvascular decompression (MVD) are exposed to the risk of permanent cranial nerve damage. Intraoperative neurophysiological monitoring of cranial nerves provides a direct and immediate feedback about the functioning of neural structures that may be inadvertently damaged during surgery, and it minimizes possible further damage [2, 4–6, 8].

This review describes our experience of intraoperative electrophysiological monitoring of the V, VII and VIII cranial nerves.

General Procedures and Technical Considerations

Patients usually undergo a preoperative neurophysiological assessment to obtain complete and detailed baseline data. This involves the recording of brainstem auditory evoked potentials (BAEP) and an audiological evaluation. For the electromyographic (EMG) study of the facial and trigeminal nerves, we use surface electrodes to record compound muscle action potentials (CMAP) and the blink-reflex. A two-channel recording machine (such as the one we use) is usually sufficient in cases scheduled for microvascular decompression or removal of small tumors.

1. Brainstem Auditory Evoked Potentials

Recordings are made after anesthesia in order to evaluate any change induced by drugs and other non-pathological factors (e.g. temperature, blood pressure and chemistry). The stimuli consist of 17.7 Hz alternating clicks delivered into the external auditory meatus at supramaximal intensity by means of earphones. The unstimulated ear is masked using white noise. Traces of 500–1500 usually unfiltered averaged responses are obtained with band pass filter at 200–3000 Hz.

2. Electromyography

Monopolar needle electrodes are inserted into the belly of the orbicularis oculis or nasalis for the facial nerves, or in the ipsilateral masseter for the trigeminal nerve. Both spontaneous activity and CMAP evoked by direct nerve stimulation are monitored.

Specific Procedures

1. Acoustic Neurinomas and Other Ponto-Cerebellar-Angle Tumors

We use both ipsilateral and contralateral BAEP recordings and EMG. The recordings made during ipsilateral stimulation are used to evaluate the integrity of the acoustic nerve during tumor removal; the contralateral side is stimulated whenever it is necessary to evaluate brainstem function or to check whether or not ipsilateral changes are due to artifacts.

2. Microvascular Decompression

In patients undergoing surgery for hemifacial spasm and trigeminal neuralgia both EMG and BAEP are recorded. Other procedures, such as recording of the direct acoustic nerve compound action potential (CAP) [7], electrocochleography or optical pathway neurophysiological evaluation are not routinely performed.

Data Interpretation

1.1. Changes in Latency and Amplitude of Evoked Potentials

In order to quantify the stability of the electrophysiological signals during surgery, we tried to evalu-

Table 1. *Facial Nerve Results.* Review of 119 patients with acoustic neuroma operated on between 1985 and 1990

Facial nerve	Monitored patients	Non-monitored patients
Identified and isolated	47/49	51/70
Unidentified or cut	1/49	12/70
Damaged and poorly responsive to electrical stimuli	1/49	7/70

ate the effect of a number of non-pathological factors which may affect evoked responses under general anesthesia.

Comparison of baseline and intraoperative data show that the amplitude and latency of BAEPs are only minimally affected; they are in fact remarkably stable during both MVD surgery and in other types of surgery performed with the patient in a sitting position.

During and after the opening of the dura mater, we have frequently observed variations in amplitude or latency of 0.3–0.4 ms. The transitory synchronization of the waveform probably reflects a change in intracranial pressure affecting the brainstem [16], and the increased latency is probably due to local physiological changes (e.g. variations in temperature caused by frequent washings of the operative field [3,5] or the slight stretching of the acoustic nerve during cerebellar retraction [13].

In view of the postoperative preservation of hearing (Table 3) [10], we consider these changes in BAEPs as not being predictive of acoustic nerve damage: in only one of four patients (undergoing MVD surgery) a mild pan-tonal hearing deficit was present.

Greater changes in latency (from 0.5 to 1.8 ms) and amplitude have been observed in the low-risk phases of an operation, such as during electric drill craniectomy in close proximity to the mastoid and, particularly,

Table 2. *Brainstem Auditory Evoked Potentials and Hearing Function in 119 Patients Evaluated 1985–1989*

Preoperative BAEP change	Usefulness of monitoring	outcome in	out
Wave abolition 78/119	none	deafness	deafness
Only waves I and II recordable: 9/119	none	deafness or severe hearing loss	deafness
Only V wave recordable: 9/119	satisfactory: 3/5 (4 not monitored)	severe hearing loss	deafness = 3 satisfactory hearing = 1 poor hearing = 1
Interwave I–III increased: 16/119	good: 10/10 (6 not monitored)	mild hearing loss	deafness = 8 satisfactory hearing = 2
Normal: 2/119	ideal: 1/1 (1 not monitored)	usually normal	mild hearing loss = 1

Comments:
– 70/119 patients did not undergo intraoperative recordings;
– only in 1/70 non-monitored patients (with intracanalar tumor) was postoperative hearing function satisfactory;
– in 21% of the 119 interventions the acoustic nerve appeared to be anatomically preserved.

Table 3. *Correlations Between Intraoperative Changes and Postoperative Outcome in 41 Patients Undergoing for Trigeminal Neuralgia*

Intraoperative changes		Hearing function
17 pts.	only minimal changes in 1 patient	moderate hearing loss
18 pts.	increase in latency and/or reduction in amplitude[a]	9: no deficit 5: only BAEP changes 2: BAEP and audiometric impairment 2: only audiometric defect
6 pts.	marked abnormalities and/or wave suppression	4: severe hearing loss or deafness 1: moderate hearing loss 1: no defects[b]

[a] No correlations have been found between changes in latency or amplitude and hearing loss.
[b] The patient had the suppression of waves during electric drill bone demolition.

during the retraction of the cerebellar lobe and the positioning of retractors. These changes have been observed more frequently and for a longer period of time during the more critical phases of surgical manoeuvres directly involving the nerve and vessels. The finding of such frequent and sometimes dramatic changes in BAEPs during the preliminary phases of surgery makes it possible that the severe hearing loss experienced by some patients operated on before the introduction of an intraoperative monitoring may actually have been caused by what were originally believed to be no-risk procedures.

It has been found that these changes in amplitude and latency are usually reversible if promptly reported to the surgeon [3,6,11,14] and that, unless subsequent modifications occur, the incidence of hearing loss is reduced. An incidence of hearing loss of about 4–5% after surgery for trigeminal neuralgia, and up to 10–12% in hemifacial spasm surgery is reported in the literature as well as in our cases [1,3]. In our 2/48 cases of hearing loss following MVD surgery and 2/49 cases of deafness following acoustic neurinoma removal significant changes in BAEPs were mainly observed during the preliminary phases of the operations.

Surgical manoeuvres performed near, or directly on the acoustic nerve produced significant BAEP changes in all of our acoustic neurinoma patients, and in many undergoing MVD; however, no clear correlation was found between the increase in latency, the reduction in amplitude and the degree of hearing loss.

The improvement in BAEPs following transient changes has recently been investigated by Moeller et al. [7], who simultaneously recorded BAEPs and acoustic nerve action potentials [6,7,13].

1.2. Wave Suppression

Complete BAEP wave suppression has been observed in all of our patients undergoing surgery for acoustic neurinomas, and in some operated on for MVD (Table 3). The suppression of all BAEP waves (regardless of whether this occurs gradually or suddenly) has usually been associated with severe hearing loss or deafness and, this association has been found even when the acoustic nerve appeared to be anatomically preserved [3,16] (Table 2). Pan-tonal hearing loss occurred only in the immediate postoperative period and it is possible that it was due to the effect of bony vibration on the transmission apparatus or the cochlea.

The abolition of BAEP wave V, with a preserved wave I, may be associated with good postoperative

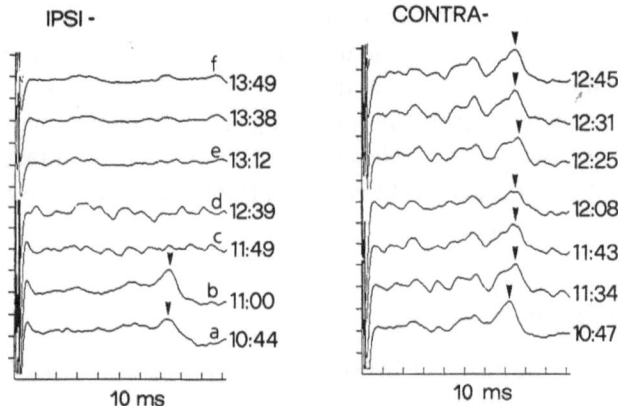

Fig. 1

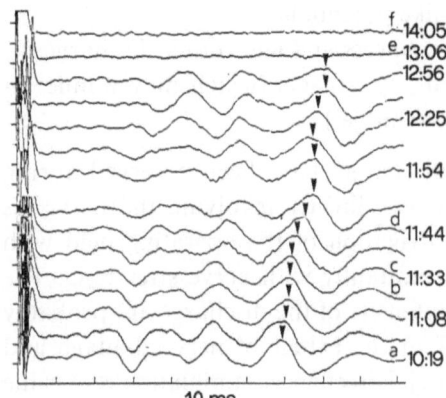

Fig. 2

hearing function. We have observed the reappearance of wave V soon after surgery in some patients, but we have never found that a suppressed BAEP recovers at the end of surgery.

If wave I is clearly present and well synchronized at the start of surgery, it should be carefully monitored because any changes in its amplitude or latency may indicate cochlear dysfunction, which may be predictive of a severe and usually irreversible loss of hearing.

In patients undergoing acoustic neurinoma surgery, BAEP wave suppression has been mainly observed during or after cerebellar lobe retraction (Fig. 1), during the dissection of the tumor capsule from the acoustic/facial bundle, and during the removal of the intrameatal portion (Fig. 2). In patients undergoing MVD surgery, the suppression has usually occurred during or after manoeuvres directly involving the acoustic/facial bundle. On the basis of the patterns of the BAEP changes, we believe that there are two mechanisms underlying hearing loss: one directly involving the nerve itself (excessive stretching, nearby coagulation and the partial resection of nerve fibers), and an indirect mechanism probably due to ischemic damage to the cochlea. Intraoperative BAEP monitoring alone is probably not enough to distinguish these two mechanisms. Therefore, methods such as direct CAP recordings and electrocochleography could prove to be useful [6,7,13].

2. Electromyography

On the basis of EMG on-line recordings and CMAP evoked by direct nerve stimulation, the following features of EMG activity during surgery may be relevant:

(a) fibrillary potentials;
(b) motor unit-like potentials:
(c) repetitive discharges and neurotonic discharges;
(d) various kinds of artifacts (which may mimic bioelectrical discharges).

Neurotonic discharges have been observed during surgical manœuvres directly involving the nerve, as well as during irrigation of the operating field with saline and concomitantly with nerve stretching. The frequency and intensity of the discharge are probably directly related to the extent of nerve damage and functional loss [2]. Direct nerve stimulation is of value because it makes it possible:

– to distinguish nevers from other structures;
– and to check their axonal integrity.

A reduction in CMAP amplitude probably indicates a partial nerve conduction block and suggests the presence of neurapraxia or structural damage (axonotmesis or neuronotmesis). During MVD surgery, only a few of our patients (with mild and transitory facial deficits) have shown changes in CMAP, none of which was significant; however, a reduction in amplitude or the complete suppression of CMAP has frequently been observed in patients undergoing acoustic neurinoma surgery, and this has invariably been associated with facial paresis or paralysis.

Conclusion

Function monitoring, first introduced in the 1950s, is now considered indispensable during all brain surgery, and the application of neurophysiological recording techniques have completely changed the scene in operating theatres over the last few years.

For surgery of acoustic neurinomas and other PCA tumors, we have found EMG monitoring to be a valuable tool in reducing the risk of nerve damage, and it should be used in all patients. BAEP recordings are useful for preserving hearing function (in patients with small tumors and good preoperative hearing), but the risk of producing a hearing loss is still high; the usefulness of this and other monitoring techniques for the preservation of hearing function in the removal of medium-sized and large tumors is generally poor. In MVD surgery, the monitoring of evoked potentials and EMG are both important means in the assessment of cranial nerve function and should be performed in all patients. Since hearing loss is the most severe complication, monitoring of BAEPs should, if possible, be supplemented by recordings of nerve action potential from the acoustic nerve [6].

Acknowledgements

We thank Prof. Franco Pluchino for his contribution of data and participation in the neurophysiological monitoring during acoustic neuroma surgery, Prof. Giuliano Avanzini for valuable suggestions concerning the manuscript and the Associazione "Paolo Zorzi" for financial support.

References

1. Auger RG, Piepgras DG, Laws ER (1986) Hemifacial spasm: results of microvascular decompression in 5.4 patients. Mayo Clin Proc 61: 640–644
2. Daube JA, Harper CM (1989) Surgical monitoring of cranial and peripheral nerves. In: Desmedt J (ed) Neuromonitoring in surgery. Elsevier, Amsterdam, pp 115–138
3. Fischer C (1988) Brainstem auditory evoked potential (BAEP) monitoring in posterior fossa surgery. In: Desmedt J (ed) Neuromonitoring in surgery. Elsevier, Amsterdam, pp 191–207
4. Friedman WA, Kaplan BJ, Gravenstein D, Rhoton AL (1985) Intraoperative brainstem auditory potentials during posterior fossa microvascular decompression. J Neurosurg 62: 552–557
5. Grundy BL, Jannetta P, Phyllis T, Procopio BA, Lina A, Boston R, Doyle E (1982) Intraoperative monitoring of brainstem auditory evoked potentials. J Neurosurg 57: 647–681
6. Möller AR (1989) Intraoperative electrophysiological monitoring during microvascular decompression of cranial nerves V, VII and VIII. In: Desmedt J (ed) Neuromonitoring in surgery. Elsevier, Amsterdam, pp 209–218 ↩
7. Möller AR, Jannetta PJ (1983) Monitoring auditory functions during cranial nerve microvascular decompression operations by direct recording from the eight nerve. J Neurosurg 59: 493–499
8. Nuwer M (1986) Evoked potentials monitoring in the operating room. Raven, New York, pp 149–171
9. Ojemann RG, Levine RA, Montgomery WW, McGaffigan P (1984) Use of intraoperative auditory evoked potentials to preserve hearing in unilateral acoustic neuroma removal. J Neurosurg 61: 938–948
10. Radtke RA, Erwin CW (1988) Intraoperative monitoring of auditory and brainstem function. In: Gilmore R (ed) Evoked potentials. Neurologic clinics series, Vol 6. Saunders, Philadelphia, pp 899–915

11. Radtke RA, Erwin CW, Wilkins RH (1980) Intraoperative BAEPs: significant decrease in postoperative auditory deficit. Neurology 39: 187–191
12. Raudzens P, Shetter AG (1982) Intraoperative monitoring of brainstem auditory evoked potentials. J Neurosurg 57: 341–348
13. Sekiya T, Möller AR, Jannetta PJ (1986) Pathophysiological mechanisms of intraoperative and postoperative hearing deficits in cerebellar angle surgery. An experimental study. Acta Neurochir (Wien) 81: 142–151
14. Scaioli V, Brock S, Ciano C, Palazzini E, Broggi G (1994) The treatment of trigeminal neuralgia by microvascular decompression. Perioperative neurophysiological evaluation. Neurology 241 [Suppl 1]: S155
15. Scaioli V, Pluchino F, Giombini S, Cella G, Franceschetti S, Avanzini G (1992) Electrophysiological monitoring during acoustic neuroma surgery. Electroenceph Clin Neurophysiol 82: 60P
16. Soulier MJ, Lazorthes Y, Fraysse B, Vincent M (1985) The monitoring of the brainstem and the auditory function in neurosurgery. In: Pluchino F, Broggi G (eds) Advanced technology in neurosurgery. Springer, Berlin Heidelberg New York Tokyo, pp 85–100

Correspondence: Giovanni Broggi, M.D., Department of Neurosurgery, Istituto Nazionale Neurologico "C. Besta", Via Celoria, 11, 1-20133 Italy.

Acta Neurochir (1995) [Suppl] 64: 40–44

A Frameless Stereotaxic Localisation System Using MRI, CT and DSA

J. Rousseau[1], D. Gibon[1], E. Coste[1,2], S. Blond[3], B. Pertuzon[4], B. Coche[5], C. Vasseur[2], and X. Marchandise[1]

[1]CLARC, Centre Hospitalier et Universitaire, [2]Centre d'Automatique, Université des Sciences et Technologies, [3]Service deNeurochirurgie, Centre Hospitalier et Universitaire, [4]Service de Neuroradiologie, Centre Hospitalier et Universitaire, and [5]Service deRadiothérpie, Centre O. Lambert, Lille, France

Summary

The authors present a method of stereotaxic localisation using magnetic resonance imaging (MRI) computerized tomography (CT) and digital subtracted angiography (DSA) which does not require localisation frams fixed to the patient's skull, but uses only four cranial landmarks corresponding to the holders of the neurosurgical stereotaxic frame. The method presents no major constraints in routine examinations. The geometrical distortions of the imaging devices are corrected. Three-dimensional localisation is performed using sagittal and axial slices in MRI, axial slices in CT and only two associated frontaly and lateral views in DSA. The images data are transferred to a PC-based system. By locating the landmarks on the images, the transformation matrixes can be computed to obtain the 3D coordinates of a target in the stereotaxic space and in any imaging modality. The results obtained show the precision of the corrections and the millimetre accuracy of pin-point target localisation.

Keywords: Neurosurgical stereotaxy; 3-D localisation; medical imaging; computer assisted system.

Introduction

In conventional stereotaxic neurosurgery, brain lesions are located with the help of standard X-ray techniques. The pictures are taken in association with a stereotaxic frame, which is fixed to the patient's head by means of four pins fastened to the outer table of the skull. The film holders are placed in a plane normal to the base plane of the frame with a long focal point-to-plate distance and the X-ray images are considered faithful, full-sized and undistorted documents. The frame provides a system of reference forming the basis for biopsies and radiosurgery. However, the modern techniques of medical imaging have become absolutely essential to allow precise localisation of tumoral or vascular targets and the associated nervous structures. These techniques necessitate the use of special devices. The commonly used methods [2,7–9,11,13] consist of N-shaped locators in CT and MRI, or pin landmarks boxes in DSA, placed on the base of the surgical frame on the patient's head. These devices pose problems due to difficulty in their fixation on the head, awkwardness of the frame, discomfort experienced by the patient, and laborious examination protocols. They give some artifacts in CT and MRI, they are often so bulky that it is necessary to use the body coil in MRI, and they limit the imaging possibilities in DSA; moreover, large FOVs are necessary to observe all of the head and the localizing device resulting in a decreased image resolution.

In order to solve these problems, we present a stereotaxic localisation method suitable for MRI, CT and DSA, and designed to locate and determine target positions without using a stereotaxic frame. A computer-assisted system is also introduced. This system enables to transposed data between images obtained from different imaging modalities, MRI, CT, DSA and X-rays, on one workstation. It computes the parameters of the stereotaxic instrument settings reproducing the path of a surgical probe to the centre of a lesion and allows direct visualisation of the points at which a simulated probe trajectory intersects the image slices.

Materials and Methods

1) Neurosurgical Room and X-Ray Picture Acquisition

In the operative room, we use a neurosurgical stereotaxic frame based on the Talairach method [14] allowing biplane serial angiography without distortion and with a constant enlargement factor. The frame provides a system of reference for locating targets and for biopsies. A CCD camera is used to digitise the X-ray pictures.

2) Localisation Landmarks

Four markers are positioned in the fixing supports of the stereotaxic frame previously placed in the outer table of the patient's

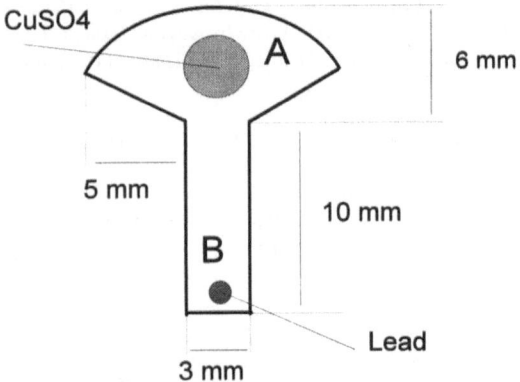

Fig. 1. Four landmarks are positioned in the place of the fixing support of the neurosurgical stereotaxic frame drilled in the outer table of the patient's skull. *A* small tank containing about 0.1 ml of $CuSO_4$. *B* small lead ball of about 1.5 mm diametre

skull. These landmarks (Fig. 1) contain at one end a 0.1 ml $CuSO_4$ solution tank for MRI localisation, and at the other end a small lead ball (1.5 mm in diameter) for CT and DSA localisation.

3) MRI Localisation

Using an 0.5 T GE MR-MAX imager, we carry out 5 mm-thick contiguous sagittal and axial slices in a gradient echo sequence (TR = 400 ms, TE = 12 ms, 256 × 256 matrixes, FOV = 250 mm) to cover the patient's entire skull, including $CoSO_4$ landmarks [10]. This procedure takes less than 10 minutes. If a tumour volume extent is still required, slices of any desired orientation can then be performed, without moving the patient's skull, with T2-weighted pulse sequences or with paramagnetic contrast agent for example.

4) CT Localisation

We use a Siemens DRH CT scanner. After sagittal scout view acquisition in order to locate the approximate plane of the fixing supports, we carry out about 10 transverse 1 mm-thick contiguous slices (256*256 matrixes, FOV = 26 cm). The lead balls of the markers are sufficient to obtain the same reference plane as in MRI.

5) DSA Localisation

A Philips DVI-S angiography system is used. Problem in DSA localisation results from the conic projection which produces an unsteady enlargement according to the position of the object in relation to the X-ray tube-detector axis. The four lead head landmarks are clearly visible on the images, and simply by knowing the real relative distances between the ends of the four head landmarks, we have shown [4] that it is possible to determine the stereotaxic localisation system and to deduce therefrom the coordinates of a target defined on antero-posterior (AP) and left-right (LR) projections. However, these distances are not known *a priori*; that is why, during a preliminary stage, we place a locating plate comprising four locating markers in a known geometrical arrangement in the immediate vicinity of the top of the patient's skull (Fig. 2). From two pictures taken in AP and LR projections, we can distinguish these four locating markers and the four head landmarks of the patient and the respective distances between the head landmarks are deduced. On pictures taken under standard clinical conditions, and in the absence of the locating plate device, only the projections of the

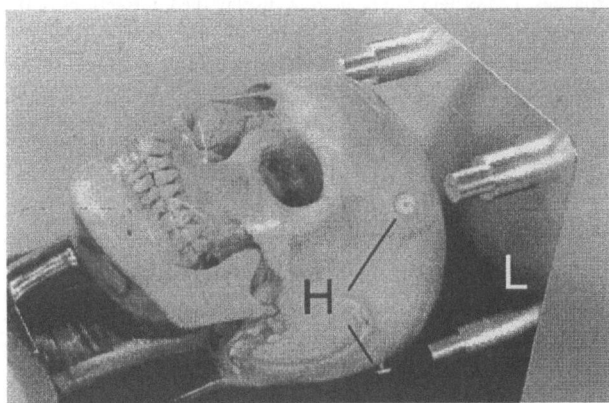

Fig. 2. Dry skull phantom on the locating plate device. *L* locating markers; *H* fiducial head landmarks

patient's four head landmarks are visible and make it possible to determine the coordinates of a target.

6) Distortions and Calibrations

Digitising by the CCD camera is dimensionally calibrated by the acquisition of a refrence circle having a known diameter. The procedure has to be performed every time the system is switched on in order to compensate for time drifts.

In MRI, 3D enlargement factors and relative offsets by which the centre of slices in one orientation is shifted from another orientation is monitored at each examination by fast acquisitions in the sagittal, axial and coronal orientations, of a cubic phantom filled with a 1g/1 solution of $CuSO_4$.

In CT, only the mechanic rigidity of the gantry and of the table gives possible misregistration. In our case, the precision of the table position is very important and has been carefully verified. Unlike teleradiography, DSA gives rise to considerable geometric distortions which depend on the orientation of the detector because the influence of the earth's magnetic field. We designed a calibration grid, and images are acquired in different orientations and FOV in order to produce matrixes for enlargement and correction of the distortion proper to each FOV and each angle of incidence.

7) Hardware and Software System

The informatic device [3] is build on a standard 486 IBM compatible computer integrating an additional image processing board, a high resolution monitor, a CCD camera to digitise the X-ray images, and magnetic tape reader to enable the MR, CT and DSA images to be transferred. The images are analysed in all modalities taking into account the pre-established calibration factors or matrixes. In MRI, the software automatically computes the 3-D coordinates of the barycentre of the markers viewed on the sagittal and axial images. In CT, the barycentres of the landmarks are obtained in a similar manner. In DSA, the user first clicks the four locating markers and the four head landmarks on the AP and LR images carried out with the locating plate; afterwards, he also clicks the four head landmarks on the acquisitions obtained without the locating plate. Then, in each imaging modality, the 3-D coordinates of the four markers are computed and are used to deduce the stereotaxic trihedron. The system can be used to calculate the 3-D position for different imaging modalities (radiography, MRI, CT, DSA), to adjust the surgical tools for orthogonal and oblique approaches. The system can be used to display the intersections of the probe with the MR or CT slices. The anatomical references, AC and PC, as well as the furthest distances from the horizontal cortical (fronto-occipital), vertical (vertex-temporal lobe) and transverse planes can be defined and are used to construct the three- dimensional proportional grid developed by Talairach, and to like up with a stereotaxic atlas. More recently, special developments make it possible to determine dosimetric treatment planning in multi-beam radiotherapy and to display isodose curvers on the images [6].

8) *Error Monitoring Procedures*

The value of the angle formed by the two diagonals of the stereotaxic trihedron should be precisely 90°; in a same manner, the distances of the markers in relation to the reference plane should be exactly 0. For each imaging modality, the determination of these values makes it possible to detect an abnormal deviation in the position of the markers, due to imaging artifact or movement of the patient, and thus assess the confidence in the calculations.

Results

To enable the accuracy of the complete localisation method to be assessed, measurements have been obtained on two phantoms. These measurements produced by our image processing software, are given in tenth of millimeters, although image pixel sizes determine the over-all attainable precision.

The first phantom consists on a 16 cm diameter PVC tube to which the 4 landmarks are inserted in the fixing positions of the stereotaxic frame. Inside this cylinder are fixed 3 simulated targets, small spherical volumes of 0.05 ml $CuSO_4$ solution. The examinations are conducted on this phantom with X-ray, MRI and CT imaging, and the targets coordinates are calculated. Table 1 gives the respective positions of the targets, measured on the images by our device. When

Table 1. *3D Stereotaxic Coordinates of the 3 Targets (First Phantom) are Measured on the X-ray, MRI and CT Images (the Target 3 was Not Viewed on X-ray Pictures)*. The similarity of values, in the order of one millimetre is comparable with the value of the pixel

Coordinates (mm)	Target 1	Target 2	Target 3
X (X-ray)	+42.5	−7.4	/
X (MRI)	+41.6	−8.5	−31.8
X (CT)	+42.4	−8.6	−32.1
Y (X-ray)	−12.9	+41.9	/
Y (MRI)	−12.4	+45.2	−30.4
Y (CT)	−11.9	+43.3	−29.1
Z (X-ray)	+67.0	+18.2	/
Z (MRI)	+68.5	+18.4	+87.4
Z (CT)	+68.3	+18.2	+88.4

Table 2. *3D Stereotaxic Coordinates of the 3 Targets (Second Phantom), Measured on Stereotaxic X-ray Images and Using CT and DSA*. The Pixel size is 1 mm for CT images and is about 0.5 mm for DSA images

Coordinates (mm)	Target 1	Target 2	Target 3
X (X-ray)	+11.5	+9.6	+5.6
X (CT)	+11.1	+8.8	+5.6
X (DSA)	+10.9	+9.4	+5.4
Y (X-ray)	+4.3	−10.9	−25.4
Y (CT)	+4.0	−11.6	−24.5
Y (DSA)	+3.8	−10.7	−26.3
Z (X-ray)	+38.2	−2.7	−44.9
Z (CT)	+36.1	−4.0	−46.0
Z (DSA)	+39.9	−2.4	−44.9

a simulated surgical operation is conducted on the phantom, simulated probe trajectories give arc adjustment parameters and each target is reached with an accuracy in the order of one millimetre, which is not exactly measurable given the respective sizes of the surgical instrument and of the volumes concerned.

The second phantom is made of a dry skull encapsulated in polyester resin. Three spherical lead targets (diameter: 0.5 mm) are placed inside a conduit running from the base of the phantom. The stereotaxic frame is positioned as in the case of a real operation. The three targets are localised in the neurosurgical operative room using X-ray pictures, CT and DSA. Table 2 gives the coordinates of the three targets of the anatomical phantom, those measured on the basis of LR and AP DSA studies compared with those obtained with CT scans and radiographs.

Discussion

1) *The Accuracy of the Method*

The accuracy of the system cannot exceed the intrinsic resolution of the imagers and it also depends on the field and matrix used (i.e. in the order of 1 mm for MRI and CT, 0.5 millimetre for DSA). In MRI, the combined use of sagittal and axial slices for the barycentre registration of the markers makes it possible to achieve an accuracy of between 1 and 2 mm on the three axes. In CT, target localisation accuracy is in the same order of magnitude in the axial slices but depends on the slice thickness in the Z-axis. In DSA, the results obtained regarding the positions of the targets demonstrate an accuracy of within one millimetre for the three coordinates when compared with those obtained in teleradiography and CT.

These results can be obtained only thanks to the corrections of the geometrical distortions of the imagers, particularly in MRI [12] and DSA. While the non-linear MRI distortions are partially rectified on most imagers by internal correction matrixes, the electronic gains of the gradients, the determine the factors of enlargement, may vary in time and a simple gain correction procedure is sufficient to provide the desired precision. The use of a cubic phantom has the advantage of permitting geometrical calibration measurements in the three planes without changing the arrangement of the phantom, and the measurements are acquired just before or just after the examination to provide the geometrical correction factors (enlargements and displacements of the slice centres) that are the most relevant at the time of the examination. In order to correct the distortion of the image intensifier in DSA, we initially attempted to use only enlargement factors established by prior calibration along the two axes of the images, but the results obtained gave errors far greater than the size of a pixel. With the use of correction matrixes, the position errors approximate by equals to the size of the pixels of the images. Here, we show only the accuracy obtained in localising pin-point targets; in the case of irregular AVM shapes, ambiguity persists as to the precise recognition of the positions and sizes of vascular targets on the basis of LR and AP projections only [1]. Three-dimensional reconstruction of the vascular tree and the use of magnetic resonance angiography [5] should make it possible to overcome these difficulties in the long term.

2) The Data Processing System

The system comprises an image data base derived from X-ray radiographs, MRI, CT and DSA images available for stereotaxic localisation. On the condition that the image files are accessable, the system can be adapted to any imaging environment. In routine use, the accuracy parameters, associated with the data acquisitions, are of fundamental importance in alerting the user to any movements of the patient during the preparatory calibration stage using the locating plate, or to any malfunction in the data processing or detection system. The adopted method does not involve image re-calculation, which necessitates the use of heavy data processing systems; on the contrary, dot by dot correction permits the use of simple image work-stations that are easier to maintain, modify and distribute, at low cost.

3) The Value of the Method

The most interesting aspect of the method resides primarily in the absence of the locating frame throughout the image examinations. This makes procedure simple as compared with conventional methods. The use of small-sized landmarks makes it possible to use a conventional MRI head coil and small FOVs in MR and CT, which guarantee good sensitivity, resolution and image quality. In DSA, the usual clinical image protocols can then be conducted without any particular constraints. Views can be taken without any inconvenience being caused by any cumbersome mechanical device attached to the patient's head. The head landmarks are sufficiently close to the patient's skull to permit the use of the customary FOVs; thus, even in examinations for stereotaxic localisation purposes, the resolution of the images remains excellent and they can be used without any risk of interfering with the radiological diagnostic work.

Conclusion

Our three-dimensional location method has proved sufficient accurate for routine clinical use. This method does not need the use of any special stereotaxic frame and the only real constraint is the installation of the landmark supports under stereotaxic conditions prior to the imaging examinations. Given its swiftness due in particular to the ease with which the patient can be positioned, the examination times are short. The proposed protocols and the quality control procedures provide millimetre accuracy and security. Our system has been in use for more than three years on a routine basis in the stereotaxic operating room and about 70 patients have undergone operations successfully. It allows a high degree of accuracy in reaching difficult targets, and it has been found to be user friendly.

Acknowledgements

This work was planned by the Centre Logistique d'Aide à la Recherche Clinique (CLARC) of the Centre Hospitalier et Universitaire of Lille (France); grants were received from the Direction de la Recherche et des Etudes Doctorales (E.A. # 1049), from the Centre Hospitalier et Universitaire (# 9301), from the Fondation pour la Recherche Médicale and from the Centre Oscar Lambret in Lille. The authors are also grateful to the GE, Philips and Siemens Companies for their valuable assistance in providing with image file organisation.

References

1. Bova FJ, Friedman WA (1991) Stereotaxic angiography: an inadequate database for radiosurgery? Int J Radiat Oncol Biol Phys 20: 891–895
2. Brown RA (1979) A stereotaxic head frame for use with CT body scanners. Invest Radiol 14: 300–304
3. Clarysse P, Gibon D, Rousseau J, Blond S, Vasseur C, Marchandise X (1991) A computer-assisted system for 3D frameless localisation in stereotaxic MRI. IEEE Trans Med Imaging 10: 523–529
4. Coste E, Rousseau J, Gibon D, Deleume JF, Blond S, Marchandise X (1993) Frameless method of stereotaxic localization with DSA. Radiology 189: 829–834
5. Ehricke HH, Schad LR, Gademann G, Wowra B, Engenhart R, Lorenz WJ (1992) Use of MR angiography for stereotactic planning. J Comput Assist Tomogr 16: 35–40
6. Gibon D, Rousseau J, Castelain B, Blond S, Vasseur C, Marchandise X (1995) Treatment planning optimization by conjugate gradients and simulated annealing methods in stereotactic radiosurgery. Int J Radiat Oncol Biol Phys, in press
7. Kelly PJ, Kall BA, Goers SJ (1986) Computer assisted stereotaxic resection of intra-axial brain neoplasms. J Neurosurg 64: 427–439
8. Leksell L, Leksell D, Schwebel J (1985) Stereotaxis and nuclear magnetic resonance. J Neurol Neurosurg Psychiatry 48: 14–18
9. Peters TM, Clark JA, Olivier A, marchand EP, Mawko G, Dieumegarde M, Muresan LV, Ethier R (1986) Integrated stereotactic imaging with CT, MR imaging and digital substracted angiography. Radiology 161: 821–826
10. Rousseau J, Clarysse P, Blond S, Gibon D, Vasseur C, Marchandise X (1991) Validation of a new accurate method for stereotaxic localisation using MRI. J Comput Assist Tomogr 15: 291–296
11. Sadler LR, Jungreis CA, Lunsford LD, Trapanotto MM (1990) Angiographic technique to precede Gamma Knife radiosurgery for intracranial arterioveinous malformations. AJNR 11: 1157–1161
12. Schad L, Lott S, Schmitt F, Sturm V, Lorenz WJ (1987) Correction of spatial distorsion in MR imaging: a prerequisite for accurate stereotaxy. J Comput Assist Tomogr 11: 499–505
13. Siddon RL, Barth NH (1987) Stereotaxic localisation of intracranial targets. In J Rad Oncology Biol Phys 13: 1241–1246
14. Talairach J, Bancaud J, Szikla G (1974) Approche nouvelle de la neurochirurgie de L'épilepsie: méthodologie stéréotaxique et résultats thérapeuriques. I. Introduction et historique. Neurochirurgie 20: 1–240

Correspondence: Jean Rousseau, PhD, CLARC-ITM, Pavillon Vancostenobel, CHRU, 59037 Lille, France.

Acta Neurochir (1995) [Suppl] 64: 45–48

3D Laser Scanning for Image Guided Stereotactic Neurosurgery

J. Taren[1], D. Ross[1], Y. Lu[2], and L. Harmon[3]

[1]Department of Neurosurgery, [2]Department of Engineering and Computer Science, and [3]Environment Research Institute of Michigan, University of Michigan, Ann Arbor, MI, U.S.A.

Summary

While this work is in its very early stages, the 3D laser scanner shows significant promise as a surgical localization device with advantages over other sensing methods. Accurate 3D surface extraction and matching, a central problem in computer vision, is the key to frameless stereotaxic neurosurgery using this technique.

Keywords: Frameless stereotaxy; 3-D laser range imaging; contour scanning; image co-registration; stereotactic neurosurgery.

Introduction

This paper describes preliminary work to develop a new approach to frameless stereotaxy in which patient-image registration is accomplished using a 3D laser range imaging sensor and precaptured CT scans. Although current and projected image guided interactive interventions rely on the scanning technologies (mechanical, optical, or acoustic) the application of 3-dimensional laser range imaging to contour scanning in neurosurgery is unique and enjoys potential advantages over these methods. The 3D laser scanner is insensitive to moisture and ambient lighting and temperature and is capable of high resolution and potentially lower computational requirements. Preliminary results support its feasibility, however, higher resolution will be necessary for clinical applications. Potential advantages include: minimal impact on surgical procedures, hands-off operation, invisible to surgeon, good standoff distance, does not affect other instrumentation, and can be made eye safe.

Concepts

A 3D imaging device is used to map the surface of the object of interest and place it in the coordinates of the operating room. By registering the surface with the same surface derived from precaptured imagery, the correspondence between image space and surgical space is obtained (Fig. 1).

3D-Laser

Three-dimensional (3D) laser scanning (laser range imaging) is being actively developed for a number of applications, including industrial parts inspection (metrology), industrial robot vision, and robotic vehicle guidance. An active sensor employs a modulated laser beam for illuminating each element of the target area and a receiver that compares the modulation phase of the laser light reflected from the target with the phase of the emitted light to determine the distance (range) to that target element. The scanner samples the target surface in a raster pattern, collecting both the range and reflectance data in a single scan. Figure 2 shows a range image in scanner coordinates (left) and cartesian (real world) coordinates (right).

The sensor operates in two modes. In the first, the working depth (ambiguity interval) is 8.4 inches, resolved into 256 steps, each measuring 0.033 in (0.8 mm). In the second mode, the ambiguity interval is 0.5 inches with range steps of 0.002 in (0.05 mm). The sensor operates at a standoff distance of approximately 36 inches, with a 35×35 degree field of view (approximately 17 inches square); the typical area resolution is 0.02 inches (0.5 mm). The scan rate is 0.5 sec per image for 320 by 320 pixels.

This type of imaging laser scanner must be distinguished from laser line scanners which produce only individual contours. The output consists of range and reflectance images in complete spatial registration at the pixel level. The gray levels in the range image

Image Space   **Surgical Space**

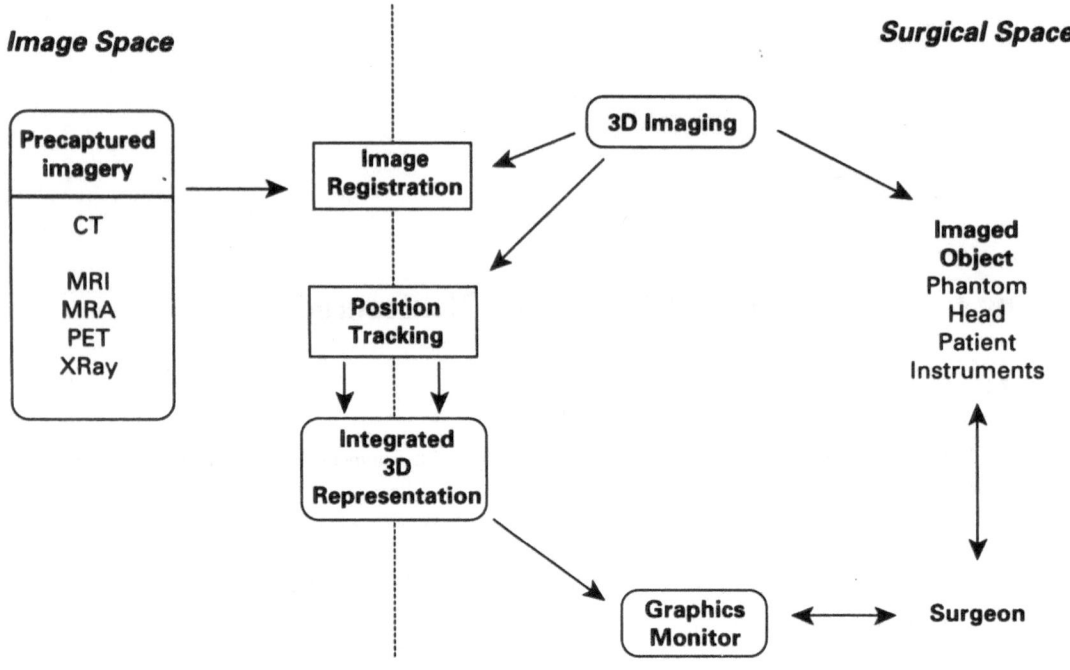

Fig. 1. 3D laser scanner localization concept

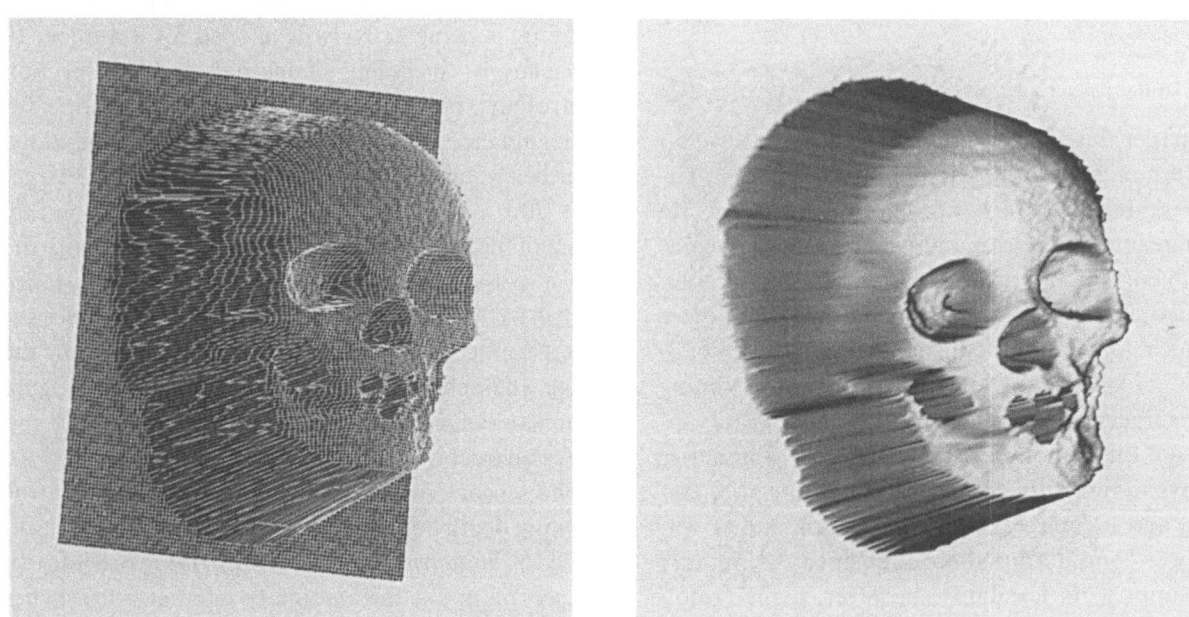

Fig. 2. Left: Range image in scanner coordinates. Right: Shaded surface rendering of range image in cartesian (real world) coordinates

represent the distance from the sensor to the target at each position in the image plane. The gray levels in a reflectance image represent the reflectance of the materials in the target at that location. Additional advantages of the 3D image laser scanner are the fast image acquisition time and frame rate. By contrast, a typical laser line scanner, which must be physically repositioned between lines, requires several seconds to acquire an image.

Laser range sensing has many advantages over infra red LED, sonic, and magnetic field sensors in the environment of an operating room: it is negligibly affected by the medium thorough which it travels; it is unaffected by and will not affect adjacent instrumenta-

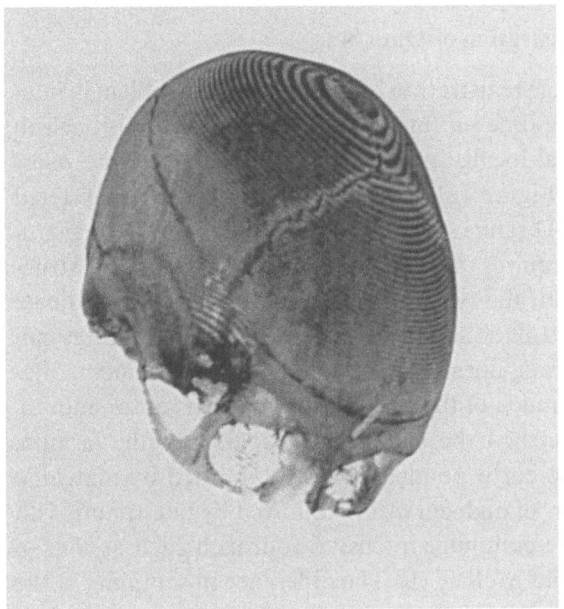 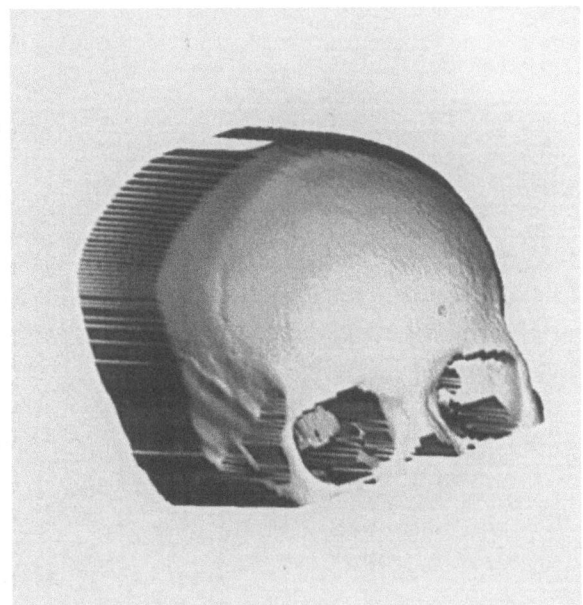

Fig. 3. Left: Volume rendering of CT data. Right: Shaded surface rendering of the surface generated by raytracing the binarized CT data

tion; and finally, ambient temperature and lighting conditions have negligible effects on sensor perform- ance. When coupled with contour-based surface- matching, no fiducial marks are required.

CT

Computed Tomography scanning is ideally suited for this project because of its high contrast resolution and high spatial resolution [2,3]. The superior contrast resolution of CT with respect to plain radiography permits direct visualization of not only the skull, but of the brain itself and also of many different kinds of pathology (brain tumors, vascular abnormalities, etc.). The constant relationship of the skull to the brain (and/or any given pathologic process inside the head) is well-demonstrated on CT.

The superior spatial resolution of CT scanning is particularly important for high precision, reproducible imaging of the bones of the skull. This sub-millimeter resolution permits intracranial structures to be repeat- edly, accurately localized without difficulty. Although Magnetic Resonance (MR) imaging does not reliably image bony structures, it provides better contrast res- olution with respect to the brain and some pathologic brain processes. MRI will therefore be considered in future studies, either alone or registered with CT.

For preliminary experiments, axial CT images of a skull were acquired with a GE 9800 CT Scanner at approximately 1 mm resolution. The 3D data sets de- rived from the range image (Fig. 2) and the CT scan are shown in Fig. 3 with binarized data (right) and a vol- ume rendering of CT data (left).

Data Registration

Our current approach to registering the surface derived from CT and laser range images is based on matching manually-selected fiducial points. The trans- formation matrix is obtained from solving the system of linear equations which arises from the set of corre- spondence points. We discuss below the extension to automatic registration of laser range and CT images, which is required for real-time localization and track- ing during cranial surgical procedures.

Automatic registration of laser range and CT data requires: 1) 2D or 3D image segmentation to extract the surface of the skull (or head) from CT images; and 2) 3D surface matching, to determine the transformation required to place the data sets in a common spatial reference system.

The external surface of the head or skull can be extracted from CT data in one of two ways: individual contours can be extracted from each slice using 2D operations and then assembled into a 3D surface; or the original slices can be assembled into a 3D data set from which the surface is extracted using 3D operations.

In our initial experiments with skulls, the extraction of the surface is relatively simple because of the high contrast of bone and the absence of other tissues. We

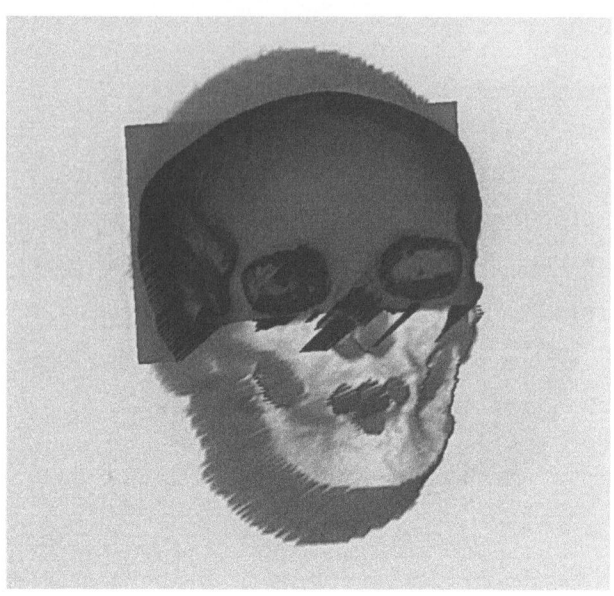

Fig. 4. CT surface (center) registered with range surface

therefore chose extract contours from individual slices. Each slice was binarized at an intensity threshold which selected bone. A connected components analysis was performed to filter out small noise and minor breaks were repaired using a morphological closing operation with a small disk. The boundaries (interior and exterior) of the skull were represented as chain codes with explicit representation of surface containment relationships. The contours from each slice were combined to form a 3D surface.

Extraction of the external surface of a head from CT requires somewhat more sophisticated processing because the surface to be registered with the range data is the external surface, or skin. However, extraction of the skin boundary is simpler than the boundaries of other soft tissues because of its location. We have had success in applying a morphological scale space approach to the extraction of tumor boundaries in CT data [4], but anticipate that simple morphological approaches will be sufficient to extract the skin surface.

Coregistration of Data Sets

Image registration is a fundamental problem in integrating different image data sets. In our approach to surgical localization, two 3D surfaces must be registered. Figure 4 shows 3D CT surface data coregistered with 3D range surface. Automatic 3D surface registration requires the identification of corresponding structures on the two surfaces, from which the coordinate transformations can be computed. The structures can be points, lines (curves), or 3D surface elements. Because much of the surface of the skull is quite smooth, we matched the surface in a coarse to fine fashion. Coarse correspondence was established by matching regions of high curvature followed by fine-tuning with a more computer-intensive approach such as that of Besl and McKay (1). The difference in sampling of the range and CT scanners means there is no a priori reason to expect individual points in the two data sets to correspond. Therefore the surface extracted from the CT data set is interpolated and modeled as a continuous surface and that from the range data set as a set of discrete surface samples.

References

1. Besl PJ, McKay ND (1992) A method for registration of 3D shapes. IEEE Trans PAMI, 14: 239, 1992
2. Boyd DP, Parker DL, Goodsitt MM (1992) Principles of computed tomography. In: Moss AA, Gamsu G, Genant HK (eds). Computed tomography of the body with magnetic resonance imaging, 2nd Ed, Vol 3. Saunders, Philadelphia, pp 1368–1372
3. Curry TS, Dowdey JE, Murray RC (eds) Christensen's physicas of diagnostic radiology, 4th Ed. Lea and Febiger, Philadelphia, pp 314–317
4. Lu Y, Harmon L (1992) Multiscale analysis of brain tumors in CT imagery. 21st applied Imagery Pattern Recognition Workshop, SPIE, October 14–16, 1992

Correspondence: James A. Taren M.D., Department of Neurosurgery, Taubman Center 2128, Box 0038, University of Michigan Medical Center, 1500 E. Medical Center Drive, Ann Arbor, Michigan 48109-0038, U.S.A.

Acta Neurochir (1995) [Suppl] 64: 49–53

Frameless Stereotaxy and InteractiveNeurosurgery with the ISG Viewing Wand

P. K. Doshi, L. Lemmieux, D. R. Fish, S. D. Shorvon, W. H. Harkness, and **D. G. T. Thomas**

Gough-Cooper Department of Neurosurgery, Institute of Neurology, London, U.K.

Summary

Preliminary experience with the ISG wand for frameless stereotaxy is presented. The wand is a multijointed mechanical arm with a range of 25 inch radius sphere. The arm has a analog/digital hybrid tansducers to locate the spatial position of the probe tip. The real time displays of triplanar 2D images with an optional 2D/3D image helps in surgical planning and interactive surgery. Seventeen patients underwent neurosurgical procedures with the help of wand. The wand was used to place minimal craniotomies, localise the brain-tumour interface and excise the lesions. The wand had an accuracy of 1.8 mm (SEM ± 0.36 mm) in localising the lesions. Thus, the use of frameless stereotaxy has made easier and more precise surgical planning in several applications.

Keywords: Computer assisted neurosurgery; frameless stereotaxy, stereotactic surgery, viewing wand.

Introduction

The conventional frame based stereotactic techniques help the surgeon to operate through minimal craniotomies, accurately localise deep seated lesions and to resect lesions from eloquent areas with confidence [8, 11, 17]. However, this involves not only invasive fixation of the skull pins but also often prolonged procedural timings to acquire intraoperative images. The fixation pins may not only be inconvenient to patient and surgeon; but may also occasionally restrict the freedom to access intracranial lesions [7, 13]. For these reasons the frame based stereotaxy is yet not widely and routinely utilised for most neurosurgical operations.

Frameless stereotaxy offers the advantage of stereotactic accuracy with a temporal and spatial freedom for surgery. Temporal freedom is achieved in terms of scanning and operating upon the patient at different timings and spatial freedom by means of getting rid of the base frame, which allows the surgeon to position and move the patient throughout the surgery. Various systems using optical [6] or magnetic [10] encorders, or ultrasonic dectectors [3] for 3D digitisation have been described. The authors present preliminary experience with a multijointed arm using proprietary analog/digital hybrid transducers for 3D digitisation. This frameless system has accuracy comparable to contemporary image guided frame systems [5, 18].

Material and Methods

The system consists of stereotactic software and hardware created by ISG technologies (Missisauga, Ontario, Canada) interfaced with the Surgicom (Faro Medical Technologies, Lake Mary, Fla.); a passive, articulated mechanical arm to perform stereotaxic spatial measurements. The arm has two degree of freedom at each of its three joints. There are two different lengths of probe which attach at the distal end of the arm. The position of the probe tip in space is calculated by the potentiometers. The position of the arm is tracked continuously and updated 30 times per second. This is interfaced with the Hewlett packard Apollo series 715 running Unix system V at 57.9 million instructions per second. The software is written in "C" and assembly languages. The presurgical workup and planning is done on ISG technology's, Allegro, 3D reconstruction workstation unit.

The detailed description of the principles and characteristics of this wand have been previously described [4, 16, 18]. Certain updates and modifications are discussed in this paper.

Preoperative Surgical Planning

One or two days prior to surgery eight reference fiducials which can be identified on the relevant imaging are stuck with adhesive patches to the patients head. A CT scan with 3 mm. thick contiguous slices and a field of view of 20–25 cm. is prescribed making sure to include the fiducials and the lesion. In the case of MRI the epilepsy scanning protocol which consists of a whole-head coronal, three dimensional volumetric spoiled gradient-echo (SPGR) acquisition with minimum TE and TR, 192 views, one repetition, 1.5 mm section thickness with 124 partitions, 24-cm field of view and 35° flip angle, is

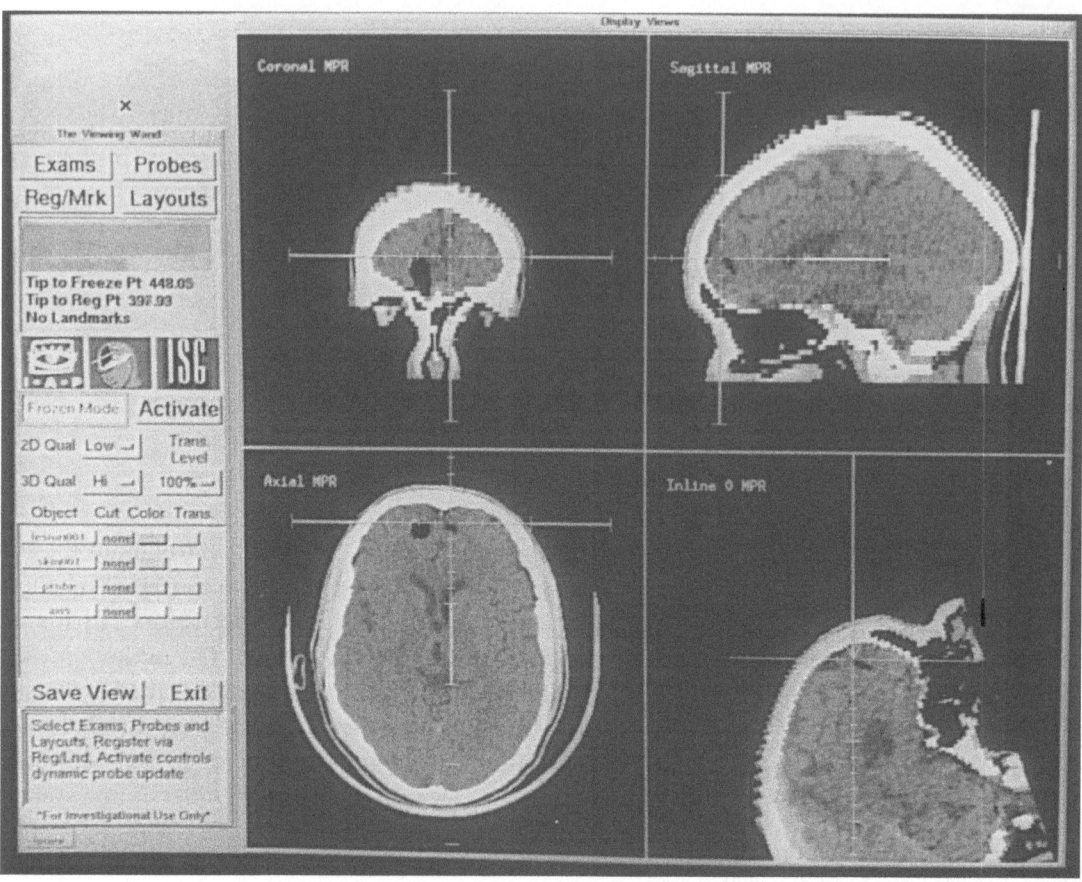

Fig. 1. The three 2D triplanar displays show the arm tip position at the meeting of the cross hair. The fourth inline view (left hand bottom) gives the reformatted view along the axis of the prode tip

prescribed. The data is then transferred either via a computer networking or on a 13 mm magnetic tape to the Allegro 3D reconstruction workstation.

A 3D model of the lesion is reconstructed using the drawing thresholding and seed growing methods of image segmentation. The skin is also reconstructed taking care to include all the fiducials in its reconstruction. From here the data is transferred on a quarter inch tape to the intheatre workstation.

Interactive Surgical Planning and Surgery

On the day of surgery, after the patient is anaesthetised he is fixed in a Mayfield head holder and the wand is clamped to the head holder. A quick check of the arm accuracy through its various joints is performed before proceeding to registration. The registration is performed by selecting four of the most accessible, least displaced and non-colinear fiducials. The registration accuracy is given by the root mean square (RMS) of the registration data set. A visual check of the registration is done by locating the probe tip position on the skin surface on all the triplanar images. The function "Tip to Landmark distance" of the computer software provides an objective assessment of checking the registration accuracy. A fiducial, not used for registration, is selected by the cursor and the probe tip is pointed onto it. The distance of the wand tip to the selected landmark is then displayed, which is the accuracy of the wand. Once registration is achieved, real time display of the probe tip in triplanar 2D views and an optional 3D/2D view is produced along the wand trajectory (Fig. 1). With the help of this information the best of the reviewed trajectories for approaching the tumour is selected. A craniotomy flap is marked appropriate for the size and site of the lesion. During the surgery the wand is used to find out the depth and position of the lesion, brain-lesion interface and to decide the extent of excision.

Seventeen patients underwent neurosurgical procedures with the wand (Table 1). In sixteen patients the wand was used to place small localised craniotomy and in fifteen cases out of these it was used interactively throughout the surgery. In one case of recurrent glioma, it was used in conjunction with CRW frame for biopsy and catheter implantation for interstitial brachytherapy

Results

System Accuracy

The maximum range of arm movements is described as a sphere of 25 inch radius. Two standard deviations of the error is ± 0.66 mm on accuracy and reproducibility. The mean RMS in all these patients was 2.15 ($\pm$ S.D. .80). The overall accuracy of the system however depends on several variables and has been found to have a mean of 1.80 mm ($\pm$ S.D. 0.36 mm) in the

Table 1. *Clinical Data of the Patients Operated with ISG Wand*

No.	Age/Sex	Lesion	Pathology	Comments
1	F/44	(R) fronto-parietal recurrent tumour	astrocytoma gr 4	wand + CRW frame
2	F/47	(R) temporo-parietal recurrent glioma	astrocytoma gr 4	craniotomy and debulking
3	F/66	(R) parietal tumour	astrocytoma gr 4	craniotomy and excision
4	F/21	(R) cingulate gyrus 1 cm sized lesion	astrocytoma gr 2	craniotomy and excision
5	F/19	(L) temporal 1.5 cm sized lesion	astrocytoma gr 1	craniotomy and excision
6	M/30	(R) frontal AVM	AVM	craniotomy and excision
7	M/36	(L) post. Temporal 1 cm lesion	astr. gr 1	craniotomy and excision
8	M/34	(L) post Temporal lesion	astr. gr 1	craniotomy and excision
9	F/35	(L) medial Temporal lesion	DNET	craniotomy and partial excision
10	M/41	(L) temporal cavernous angioma	cavernous angioma	craniotomy and excision
11	F/27	(L) Medical temporal sclerosis	hippocampal sclerosis	cramiotomy and temporal lobectomy with amygdalo-hippocampectomy
12	M/50	(R) occipital glioma	astr gr 4	craniotomy and excision
13	F/39	(R) parietal meningioma	meningioma	craniotomy
14	F/37	(R) occipital AVM	AVM	craniotomy and excision
15	F/26	(R) frontal cortical scar	cortical Scar	craniotomy and Partial excision
16	M/18	bifrontal dermoids	dermoids	craniotomy and excision
17	M/45	(L) parietal glioma	astr gr 4	craniotomy and debulking

clinical setup. This is quite comparable to the other frame based systems [4]. The more widespread the fiducials are, the better is the registration accuracy. We have compared the accuracy data of fiducial (n = 10) and surface matching (n = 7) methods of registration and have not found any statistical difference between the two.

Of the seventeen patients operated with the assistance of the wand there was no acute morbidity. The wand helped in achieving total excision of all the well

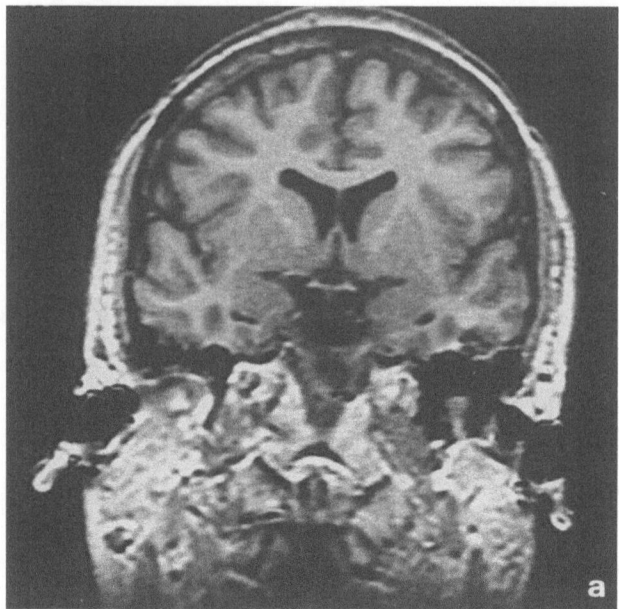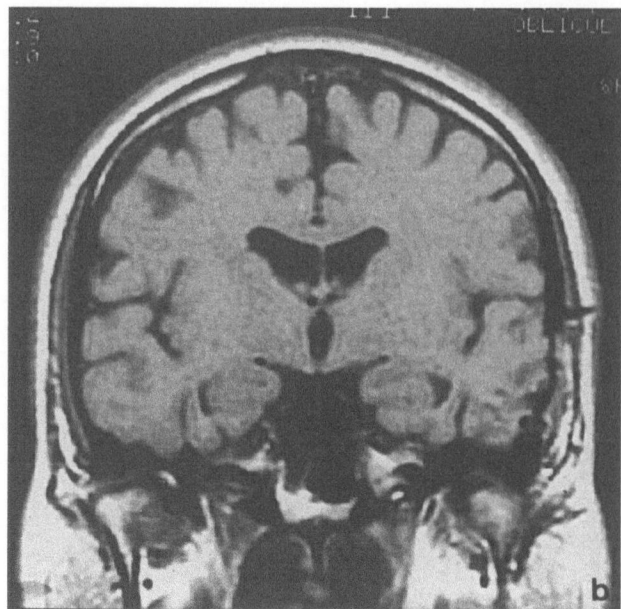

Fig. 2. Pre- and postoperative MRI of a patient having left temporal lesion

circumscribed lesions (Fig. 2). Three out of four patients with glioma improved in their clinical status after surgery and one remained stable. Of the eleven cases with lesional epilepsy eight patients had improvement (Engel gr I) in their seizures.

Discussion

From the initial description of Clarke's [8] stereotactic systems to late nineteeneighties, word 'stereotaxy' meant, use of a base ring and arc systems. With the advent of CT/MRI compatible frames stereotaxy was extended more widely to biopsies of masses [14], aspiration of abscesses [1], evacuation of haematomas [2] or insertion of radioisotopes [15]. Kelly *et al.* developed a computer graphics assisted arc quadrant system for volumetric resections of brain tumours [11, 12]. However all these systems were still based on use of a frame. The development of frameless stereotaxy depended on improved quality of neuroimaging and computer graphics.

Most of the frameless stereotactic methods described registration using one or the other types of marker fixed to the patients head during the scanning. With the ISG software it is possible to register the patient to the image data set without using any markers. This is done by visually correlating five anatomical landmarks and selecting and correlating thirty surface points. The registration achieved by this way is comparable to the fiducial registration. (Fiducial based registration (n = 10) Mean = 1.6, Surface registration (n = 7) Mean = 1.9). All our epilepsy cases are usually scanned under epilepsy scanning protocol and hence it only requires the transfer of data across the ethernet, thus obviating the need for extra scan.

The ISG wand clamps to the Mayfields headrest to which the patient's head is fixed. This allows the table position to be changed during the surgery without disturbing the relationship of the wand to the patients head. The way it attaches to the mayfield keeps it away from the surgical field thus allowing complete freedom in draping and operating, even with a microscope and eclectrocorticography. Like other frameless systems [6, 10] the encorders in the arm are not influenced by the ambient environment, instrumentation or the line of sight problems.

The tip to manual distance function gives the distance from the tip to the manually selected target. This not only assists in checking the accuracy and planning the surgery but is also very useful interactively. In case 11 the posterior extent of the tempral lobectomy on each gyrus was planned by measuring the distance from the temporal pole. Thereafter before resecting the amygdala-hippocampus complex, again the length of hippocampus to be resected was measured. This correlated well with the excised specimen.

The other function of the software is to provide 2D inline view along the wand trajectory. This view is

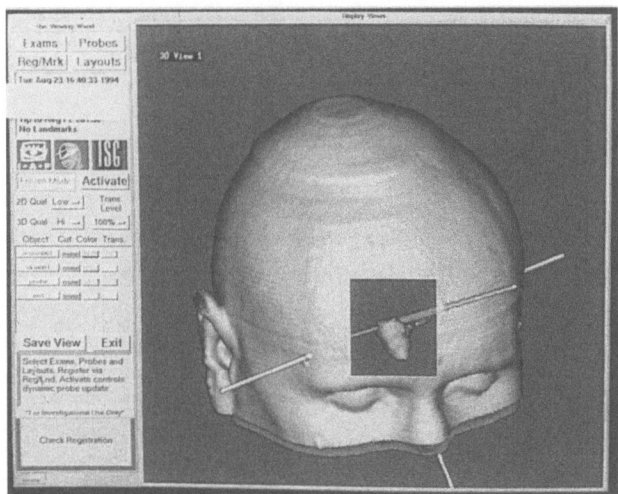

Fig. 3. The 3D view with the translucency window displays the tumor as visualised in relation to the skin surface, this is helpful in planning craniotomies

helpful in locating the brain tumour interface and deciding the extent of excision. The other functions of translucency and cut plane offers small but definite advantage in the planning and execution of the surgery (Fig. 3).

The accuracy of the wand depends on a number of variables. The preoperative factors include scanning proctocols, slice thickness, patient movement, scan resolution and image segmentation. The operative factor is the brain shift. The wand does not have capacity to update its images for the brain shift and hence, once this has occured it may be difficult to detect and compensate. Various avenues are being explored to answer this problem, these include intraoperative ultrasound, per-operative imaging and tracking the brain surface with video imaging.

The potential of this system is considerable. Its use for endoscopy, biopsy and skull base surgery has been described. It's use can be extended to radiosurgery where fractionated radiotherapy can be administered. Its utility in the spinal surgery and transoral and transsphenoidal surgery is also being investigated. The pace of progress is such that computer assisted neurosurgery of this kind is likely to become a routine procedure. The viewing wand with its potentialities and continuous upgradation will have a definite place in frameless stereotaxy.

References

1. Appuzo MLJ, Chandrasoma PT, Cohen DM, Zee CC (1987) Computed imaging stereotaxy: experience and perspective related to 500 procedures applied to brain masses. Neurosurgery 20: 930–937
2. Backlund E-O, von Holst H (1978) Controlled subtotal evacuation of intracerebral hematomas by stereotactic technique. Surg Neurol 9: 99–101
3. Barnett GH, Kormos DW, Steiner CP, Weisenberger (1993) Intraoperative localisation using an armless, frameless stereotactic wand. J Neurosurg 78: 510–514
4. Drake JM, Rutka JT, Hoffman HJ (1994) Instrumentation technique and technology: ISG viewing wand system. Neurosurgery 34: 1094–1097
5. Galloway RC, Maciunass RJ, Latimer JW (1991) The accuracies of four stereotactic frame systems: an independent assessment. Biomed Instrum Technol 25: 457–460
6. Guthrie BL, Adler JR (1993) Computer-assisted preoperative planning, interactive surgery, and frameless stereotaxy. Clin Neurosurg 37: 112–131
7. Hassenbusch SJ, Anderson JS, Pillay PK (1991) Brain tumor resection aided with markers placed using stereotaxis guided by magnetic resonance imaging and computed tomography. Neurosurgery 28: 801–806
8. Horsley V, Clarke RH (1908) The structure and functions of the cerebellum examined by a new method. Brain 31: 45–124
9. Kartimenos GP, Thomas DGT (1993) The role of image-directed biopsy in the diagnosis and management of brainstem lesions. Br J Neurosurg 7: 155–164
10. Kato A, Yoshimine T, Hayakawa T, Tomita Y, Ikeda T, Mitomo M, Harada K, Mogami H (1991) A frameless armless navigation system for computer-assisted neurosurgery. J Neurosurg 74: 845–849
11. Kelly PJ, Earnest F IV, Kall BA, Goerss SJ, Scheithauer B (1985) Surgical options for patients with deep seated brain tumors: computer assisted stereotactic biopsy. Mayo Clin Proc 60: 223–229
12. Kelly PJ (1988) Volumetric stereotactic surgical resection of intra-axial brain mass lesions. Mayo Clin Proc 63: 1186–1198
13. Moore MR, Blach PM, Elenbogen R, Gall CM, Eldredge E (1989) Stereotactic craniotomy: methods and results using the Brown-Roberts Wells stereotactic frame. Neurosurgery 25: 572–578
14. Ostertag CB, Mennel HD, Kiessling M (1980) Stereotactic biopsy of brain tumours. Surg Neurol 14: 275–283
15. Ostertag CB (1993) Interstitial irradiation treatment of low grade gliomas. In: Thomas DGT (ed) Stereotactic and image directed surgery of brain tumours. Churchill Livingstone, London, pp 125–134
16. Sandeman DR, et al (1994) Advances in image directed neurosurgery: preliminary experience with the ISG viewing wand compared with the Leksell G frame. Br J Neurosurg 8: 529–544
17. Thomas DGT, Bradford R, Bydder G (1987) Magnetic resonance directed stereotactic brain biopsy. J Neurol Neurosurg Psychiatry 50: 645–648
18. Zinreich SJ, Tebo SA, Long DM, et al (1993) Frameless stereotactic integration of CT imaging data: accuracy and initial applications. Radiology 188: 735–742

Correspondence: Paresh K. Doshi, M. D., Gough-Cooper Department of Neurosurgery, Institute of Neurology, Queen Square, London WC 1N 3BG, U.K.

Acta Neurochir (1995) [Suppl] 64: 54–58

The Impact of Interactive Image Guided Surgery: The Bristol Experience with the ISG/Elekta Viewing Wand

D. R. Sandeman and **S. S. Gill**

Department of Neurosurgery, Frenchay Hospital, Bristol, U.K.

Summary

From June 1992 to August 1994 we have accumulated a 305 case experience with the ISG Viewing Wand, the first commercially available system for interactive image guided neurosurgery. Prior to the arrival of the wand 2.5% of intracranial procedures were carried out using the Leksell G frame for image guidance. Since the arrival of the wand that percentage of procedures suitable for image guidance has increased to 10%. The wand was used for 287 supratentorial procedures, 108 craniotomies, 48 trephine exposures, 34 burr hole biopsies, 4 ventriculoscopies and 1 shunt insertion. 19 posterior fossa explorations were performed. 28 skull base procedures, including 22 transsphenoidal pituitary operations, 3 petrousectomies and 3 orbital explorations were also carried out. In addition 3 spinal cases were included in the series, 2 transoral explorations and 1 sacral laminectomy. The technique was applicable to 193 tumour cases, 14 vascular cases including 7 aneurysms and 16 epilepsy cases. Both CT (70%) and MRI (30%) scans were used for image guidance. Contour matching algorithms were used for registration throughout. Since the arrival of the wand conventional stereotaxy has been used for 26 cases, 22 stereotactic biopsies, 2 thalamotomies and 2 craniotomies (0.5% of intracranial procedures). We conclude that contour matching, interactive image guidance using a mechanical arm has replaced frame based stereotaxy in our department except for point source localisation in deep midline structures. It is a technique that has universal application to intracranial neurosurgical procedures and as such represents a major advance in image guided neurosurgery.

Keywords: Interactive image-guided surgery; frameless stereotaxy; computer-assisted surgery; robotics.

Introduction

The Frenchay Neurosurgical Department in Bristol, UK provides a comprehensive neurosurgical service to a population of 2.5 million in the South West of England, performing in excess of 2200 operations a year of which 1200–1300 are intracranial procedures. This activity has to be fed through two operating theatres and 48 adult neurosurgical beds so there is considerable pressure to make efficient use of resources in the department. In June 1992 interactive image guided surgery was introduced to the department in the form of the ISG Viewing Wand. As this as the first installation outside the umbrella of the Federal Drug Administration and the first to apply the technology to a complete cross section of intracranial neurosurgery we set out to study the use of the Wand with four main aims in mind: 1. to examine the range of application of the system, 2. to determine the application accuracy of the system, 3. to study the logistics of wand use in a busy practice such as ours and 4. to determine the universality of application or otherwise of contour matching interactive image guided neurosurgery. This paper is a summary of our first 27 months experience with the system.

Material and Methods

The ISG viewing wand is a free standing mechanical arm system for interactive image guided surgery that allows registration of a three dimensional data set to the patient using contour matching algorithms [2]. The Wand itself, first developed by the Canadian company, ISG and now refined and marketed by the Swedish company, Elekta, consists of a mechanical arm with six moveable joints each with a 320 degree range of movement. Each joint contains potentiometers that constantly relay the position of each joint and therefore the tip of the Wand to the host computer, the Hewlett Packard 710.

The ISG/Elekta Wand software is Unix based image processing software that allows accurate segmentation and seeding of the skin surface on either CT or MRI scans to produce accurate three dimensional reconstruction from two dimensional scan data [6]. The software also allows for near real time reformat of the two dimensional scan data in any plane including the standard tri-orthogonal planes and an in line view in the direction the Wand is pointing.

To be useful clinically the position of the pre-operative images on the computer have to be registered accurately to the position of the patient's head [5]. First the arm itself is attached using a special adapter to the three pin head rest. The registration process itself consists of three parts. Firstly the position of recognisable skin features are registered by placing the wand on the skin point, always the nasion and then a combination of points around the eyes and ears registering the same point on the 3D image on the computer. A minimum of three fixed points with a maximum of five points are used.

The second part of the registration process involves placing the Wand on a minimum of thirty randomly placed points on the scalp surface in a distribution that covers the whole circumference of the scalp. The contour matching algorithm is then run on the computer to complete the registration.

Finally and essential to the whole process of registration to preoperative images is a thorough accuracy check. First the wand is placed on the recognisable skin features and the position compared to that of the wand tip on the computer screen. The second check is to place the wand lightly on the skin in all four quadrants of the scalp and observe its position on the triorthogonal display on the computer. The Wand is then pressed in towards the bone and the degree of movement of the tip observed on the screen. A symmetrical exact match of wand position is required to be certain of an application accuracy before intraoperative brain shifts occur of $+/-2$ mm. A variety of tips can be attached to the Wand. We have used long and short straight probes, a bayonet probe, a biopsy cannula, various designs of electrode carrier and the rigid ventriculoscope.

Results

The overall series of cases to data is summarised in Tables 1–4. A total of 318 operations in 305 patients were performed, including 13 multiple procedures under the same anaesthetic. The mean registration time for the series was 21 minutes with a range of 5–70 minutes. The mean operation time was 2.32 hours (median 2 hours) with a range of 20 minutes to 13 hours. Mean length of hospital stay was 8.9 days (median 6 days) with a range of 1–154 days. This compares with an overall mean length of stay for all intracranial cases in the department of 10 days (median 7.7 days, range 1–423 days).

Table 1 summarises the experience to data with supratentorial tumour surgery. 25% of intrinsic tumours were biopsied either directly or using the ventriculoscope, 50% had standard craniotomies and 25% were suitable for a minimally invasive surgical approach and resection. The minimally invasive approach to gliomas supported by the Wand, the extra certainty of tumour margins offered by the wand and the reduction in disruption of the normal brain that

Table 1. *Supratentorial Tumour Surgery*

	Biopsy	Craniotomy	Trephine	Ventriculoscopy	Transphenoidal	Total
Intrinsic tumours						
Malignant glioma	25	56	21	1	–	103
Low grade astro	5	15	15	–	–	35
Lymphoma	3	1	1	–	–	5
Total	33	72	37	1	0	143
Extrinsic tumours						
Meningioma	–	12	3	–	–	15
Metastasis	1	12	9	–	–	22
Pineal	1	3	–	3	–	7
Pituitary	–	4	1	–	24	29
Craniopharyngioma	2	1	1	–	–	4
Total	4	32	14	3	24	77
Missed lesion	1	1	–	–	–	2
Total	37	105	52	4	24	222

Table 2. *Infratentorial Tumour Surgery*

	Craniotomy	Post fossa	SOC	Petrousectomy	Total
Meningioma	–	3	1	–	4
Cholesteatoma	–	–	–	2	2
Dermoid	1	–	1	–	2
Acoustic	–	–	3	–	3
Glomus	–	–	1	1	2
Metastasis	–	–	–	1	1
Brain stem tumour					
Low grade astrocytoma	–	2	–	–	2
Malignant glioma	–	2	–	–	2
Clivus tumour	2	–	1	–	3
Cerebellar					
Haemangioblastoma	–	1	–	–	1
Juvenile astrocytoma	–	1	–	–	1
Medulloblastoma	–	1	–	–	1
Total	3	10	7	4	24

SOC suboccipital craniectomy.

can be achieved with wand surgery is reflected in the length of stay for the different glioma operations. The mean length of stay for both surgical biopsy and for definitive resection was 6.1 days. Of the surgical resection cases, for those undergoing standard craniotomy the mean length of stay was 6.7 days. For trephine and minicraniotomy cases it was 4.9 days.

The system has been used in the surgery of 77 extrinsic tumours. For metastases and some meningiomas the wand was used in the same way as for intrinsic tumours, i.e. to locate and resect lesions within the brain with the minimum disruption to the normal brain. With skull base meningiomas the wand was used for orientation within the tumour and for locating important structures adjacent to or within the tumour. For pineal tumours the wand was used either to aid orientation during direct surgery or to orientate the ventriculoscope during ventriculoscopic biopsy and third ventriculoscopy. In trans-sphenoidal approaches to the pituitary gland, the wand was used to provide tri-orthogonal orientation during the approach, an advantage over our standard technique of lateral X ray screening alone. This allowed the sphenoid fenestration to be made in precisely the right place so that fracture of the Vomer could often be avoided.

In two patients (< 0.5%) the lesion was missed and histology not obtained. This compares favourably with series of other forms of image guided surgery [7].

The wand was easily applicable to all types of posterior fossa surgery included cerebello-pontine angle tumours, intracerebellar tumours and brain stem tumours. However the series only contains three acoustic neuromas out of a total of 75 acoustics operated on in the department during the study period. The acoustic surgeons very quickly decided that the wand did not enhance their already minimally invasive sub-occipital approach as it was not possible to image the anatomy that they wished to avoid.

Tables 3 and 4 summarise the use of the wand for indications other than tumour surgery. All the non lesional epilepsy surgery in the department is now carried out wand directed. Rather than use anatomical cues to orientate the dissection for a selective temporal lobectomy, a trans-sulcal approach is now used having first used the wand to place deep brain electrodes for peroperative recording. The extent of resection is now tailored to the peroperative neurophysiology. The exact extent of a callosotomy can easily be determined by the wand and the placement of deep brain electrodes can be carried wand directed much more quickly than with frame based techniques.

One application for the wand that has extended the range of image guided surgery is in vascular surgery. We started using the wand to locate peripherally placed aneurysms that were difficult to find with conventional techniques. However the use of the wand has increasingly been extended to include more conventionally placed aneurysms e.g. pericallosal and middle cerebral aneurysms, where the wand allows a minimally invasive approach to be carried out to identify the proximal vessel and the aneurysm safely without extensive anatomical dissection.

Table 4 outlines the other non tumour applications of the wand. The role in abscess surgery is the same as

Table 3. *Non Tumour Surgery – Non Lesional Epilepsy*

	Biopsy	Craniotomy	Trephine	Total
Selective temporal lobectomy	–	–	16	16
Deep brain electrodes	–	–	7	7
Callosotomy	–	2	–	2
Multiple subpial transection	–	1	–	1
Neurophysiology	–	1	–	1
Total	0	4	23	27
Vascular surgery	Craniotomy	Trephine	Post fossa	Total
Aneurysm	4	3	1	8
AVM	4	–	3	7
Cavernoma	3	1	4	4
Intracerebral haematoma	–	1	–	1
Total	11	5	4	20

Table 4. *Non Tumour Surgery – Other*

	Biopsy	Burrhole	Trephine	Crany	Post fossa	Orbity	T/oral	Laminy	Total
Abscess surgery	–	7	4	–	1	–	–	–	12
Orbital surgery	3	–	–	4	–	2	–	–	9
Hydrocephalus	–	1	–	–	–	–	–	–	1
Spinal surgery	–	–	–	–	–	–	2	1	3

Crany craniotomy, *Orbity* orbitotomy, *T/oral* transoral, *Laminy* laminectomy.

for tumour surgery. Of particular value is the ease with which multiple abscess can be drained. Orbital surgery has been greatly enhanced by image guided techniques, particularly the biopsy of orbital lesions. In one case of hydrocephalus associated with a Russell Pennybacker cyst. The wand was used to locate and drain both the ventricle and an isolated cyst with the same catheter.

We have started to explore the use of the wand in the spine. High cervical surgery can be carried out image directed using exactly the same technique as for skull base procedures. Application for surgery to the rest of the spine however is limited by the additional problems of registration when the wand base cannot be fixed firmly to the relevant anatomy. We have overcome this in one case of a sacral osteoblastoma by scanning the patient in the operative position, fixing the wand to the operating table and minimising the movement between the patient and the wand base. We then registered the image to bony landmarks on the spine at the same point in expiration and were able to achieve an accuracy of $+/-5\,\text{mm}$, sufficient to localise a $2\,\text{cm}$ diameter osteoblastoma. However the problems of registration will limit the universal application of the technique in the spine.

Discussion

Frame based stereotaxy has never been universally applicable to general intracranial neurosurgery [5]. Reasons for this include problems of accessibility with the frame in place, particularly in the posterior fossa and around the skull base, the difficult and time consuming logistics of peroperative scanning and the reluctance of general neurosurgeons to learn stereotactic techniques. Use of the viewing wand is applicable in any intracranial case where a pointer that relates intraoperative position to a preoperative image to an accuracy of no more than $+/-2\,\text{mm}$ has advantages. The range of application will therefore vary with an individual surgeon's requirements, but as this study shows, the system can have application to a complete cross section of intracranial neurosurgery, enhancing minimally invasive surgical techniques, reducing the exposure required, minimising trauma to the normal brain and aiding localisation of vital structures in anatomy distorted by the pathology. The system has extended the range of image guided surgery to include the posterior fossa and brain stem, the skull base and the orbit and is applicable to pathology not usually associated with image guided techniques. e.g. vascular surgery and pituitary tumours. The indications for use of the system in the spine, apart from the high cervical region will prove limited because of the different registration technique required when compared with good real time biplanar X ray screening.

The implications of a universal enhancement of minimally invasive neurosurgery will need to be carefully assessed not only in terms of outcome and side effects of surgery, but also in terms of treatment protocols, health economics and training. Already we have seen a greater reliance on radical surgery in our treatment policies. We have witnessed a tangible benefit in terms of theatre utilisation and patient throughout in the fact that we have sustained a 2.5% increased workload on a 10% reduction in available beds. The implications for the training of junior neurosurgeons will be enormous as the effects of European legislation take effect in the UK, effectively halving the length of time a junior surgeon can now spend in a training grade.

The use of contour matching is unique to the viewing wand and is of fundamental importance as it is this allows registration to preoperative images. Universal availability of images for interactive image guided surgery is not possible if a fiducial scan is required before each intervention. The theoretical advantage of fiducial based systems in terms of accuracy is nullified by the way in which interactive image guided systems are used, as an adjunct to craniotomy where an application accuracy of greater than $+/-2\,mm$ is neither possible nor necessary [3].

Lack of real time update with the system is always presented as a theoretical objection to its applicability. However the system is interactive in that it allows the surgeon to constantly update the registration to both fixed and moveable anatomical structures simply by placing the wand on a structure and assessing its position on the computer screen. It is this fundamental difference to frame based stereotaxy combined with the universal application of the wand and systems like it that make the term "frameless stereotaxy" such a misnomer. We recommend that the term "Interactive image guided neurosurgery" coined by Macuinas should be adopted instead [4].

What is the role of frame based stereotaxy in the future? Some light has been thrown on this question by this series. The Leksell G frame has continued to be used for the point localisation of deep brain structures in procedures such as thalamotomy and thalamic biopsy. We perceive that the stereotactic frame will persist in this its traditional role until such time as lockable arms have been fully developed. The relocatable stereotactic frame has created the ability to fractionate stereotactic radiosurgery and as a result the main developments for frame use will be in the delivery of all head, neck and brain radiotherapy [1].

The technology of the location device in interactive image guided surgery is developing rapidly. Mechanical arms will be replaced by armless systems. The focal length of a microscope can now be registered to the preoperative image. Lockable arms are being developed which will have applications, particularly in neuro-endoscopy. Active robotic arms are also a possibility for the future but will present a completely new set of problems to overcome. However the major breakthrough in this field bas been made with the ability to register a patient to a three dimensional data set. Modern neurosurgery began with the advent of cross sectional imaging and the microscope. Interactive image guided neurosurgery is technology that bridges the gap between the two and as such represents as fundamental an advance.

References

1. Gill SS, Thomas DGT, Warrington AP, et al (1991) Relocatable frame for stereotactic external beam radiotherapy. Int J Radiat Oncol Biol Physics 20: 599–603
2. Leggett WB, Greenberg MM, Gannon WE, et al (1991) The viewing wand: a new system for three dimensional computed tomography correlated intraoperative localisation. Curr Surg 48: 674–678
3. Macuinas RJ, Galloway RL, Fitzpatrick JM, et al (1992) A universal system for interactive image directed neurosurgery. Stereotact Funct Neurosurg 58: 108–113
4. Macuinas RJ (1993) Interactive image-guided neurosurgery. Neurosurgical topics. Abstract. Meeting of American Ass Neurol Surg
5. Sandeman DR, Patel N, Chandler C, et al (1994) Advances in image-directed neurosurgery: preliminary experience with the ISG viewing wand compared with the Leksell G frame. Br J Neurosurg 8: 529–544
6. Sandeman DR, Case A, Cause E (1992) Frameless stereotaxy: a new era in image directed neurosurgery and radiological image processing. Clin MRI 2: 91–92
7. Thomas DGT, Nouby RM (1989) Experience in 300 cases of CT-directed Stereotactic surgery for lesion biopsy and aspiration of haematoma. Br J Neurosurg 3: 321–326

Correspondence: David R. Sandeman, Consultant Neurosurgeon, Frenchay Hospital, Bristol, Avon UK BS16 1LE, U.K.

Acta Neurochir (1995) [Suppl] 64: 59–63

Stereotactic Endoscopic Interventions in Cystic Brain Lesions

D. Hellwig[1], B. L. Bauer[1], and E. List-Hellwig[2]

Departments of [1]Neurosurgery and [2]Radiology, Philipps University Marburg, Marburg, Federal Republic of Germany

Summary

Stereotactic endoscopic techniques are extremely helpful in diagnosis and therapy of cystic intracerebral space occupying lesions. Acute space occupying lesions can be managed effectively and without major tissue traumatization. Up to now we have operated on more than 70 cystic intracerebral space occupying lesions with a stereotactic endoscopic technique. The main diagnoses were colloid cysts, cystic craniopharyngeoma, arachnoidal and pineal cysts. In must be stressed that in cystic anaplastic astrocytomas and glioblastomas as well as metastases only an acute inner cerebral decompression can be achieved by neuroendoscopic techniques in combination with the application of reservoir systems. In benign parenchymal or intraventricular cysts neuroendoscopic intervention is performed for definitive treatment. The results are overall encouraging. There was no operative mortality and operative morbidity was below 3%. Postoperative follow-up in patients with benign cysts showed no evidence of recurrence.

Keywords: Stereotactic neurosurgery; neuroendoscopy; cystic brain tumour; colloid cyst; craniopharyngioma; pineal cyst.

Introduction

Intracerebral cystic lesions are a domain for stereotactic endoscopic interventions [3,5,12]. This group of pathological space occupying cavities includes colloid cysts, cystic craniopharyngiomas, pineal cysts, arachnoideal cysts, as well as malignant tumours with cystic components as it is the case with anaplastic astrocytomas, glioblastomas and metastases. Cystic lesions of the third ventricle have to be discussed separately. From an anatomic and functional point of view the third ventricle can be divided in three compartments: the anterior, the posterior, and the superior part. Raimondi describes a fourth parasellar part [19].

Microsurgical interventions on cystic lesions, located in the different parts of the third ventricle, demand various operative approaches [7,20,22,24,25]. They carry a high risk of complications. In contrast, with a frontal, precoronar stereotactic endoscopic burrhole approach all compartments of the third ventricle can be reached easily with a flexible steerable endoscope and the cystic lesions can be evacuated in minimal invasive, less traumatical technique. Colloid cysts located in the anterior and superior part, cystic craniopharyngiomas in the parasellar and anterior part, and pineal cysts in the dorsal part of the third ventricle are typical indications for stereotactic-endoscopic interventions.

Material and Methods

Instruments

Different rigid and flexible endoscopes for neuroendoscopic interventions are presently available. The specifications vary and one must choose according to the planned operative application. For neuroendoscopic interventions in intracerebral, space occupying cysts we prefer the flexible steerable endoscope together with the rigid Marburg Endoscopy Fixation and Guiding System which is compatible to our stereotactic system. The system consists of a special self-retaining arm, which provides the necessary stability during the operative procedure. The endoscopic working depth is regulated by micrometer screws. The metallic guiding system consists of bougies and guiding tubes of different diameters and length. Supplementary instruments are microforceps for biopsy, microscissors for cyst fenestration, bare laser fibers for hemostatic procedures and vaporization and radiofrequency probes. Digital dynamic subtraction radiography is used intraoperatively for ventriculography or cystography.

Operative Technique

Endoscopic interventions on cystic intracerebral processes can be performed "free-hand" or with stereotactic technique. We prefer the stereotactic-endoscopic technique because it offers the advantage of CT or MRI precalculated guidance to the region of interest [12,13]. It is advisable to choose two target points, one at the outer cyst wall and one at the bottom of the cyst. Both can be reached using one trajectory without changing angles at the stereotactic system. After burrhole trepanation, incision of the dura mater and coagulation of the arachnoidea, the endoscopy guiding tube is inserted and guided to the calculated target point (outer surface of the cyst wall). The inner guiding tube is removed and the cyst membrane can be inspected. In inventricular cystic processes (colloid cyst, pineal cyst) the anatomic-topographical relationships can be judged with good optical quality, whereas in intraparenchymal cysts rinsing of saline solution is necessary to get an impression about the surface of the cyst. Endoscopic inspection and judgement of the capsule consistence and its vascularization is an obvious advantage against "blind" stereotactic evacuation of cystic lesions, because the main problem with the perforation of elastic cyst walls is the occurrence of post-punctional hemorrhage. The opening of the cyst membrane is performed with the help of microscissors or microcatheters. Thin membranes can be fenestrated by use of radiofrequency or laser coagulation. After fenestration the endoscope is guided through the cyst to the second calculated target point. The cyst contents are aspirated either directly through the endoscope's working channel or through an applied microcatheter. If the contents is too viscous (colloid cyst, cystic craniopharyngioma) it is aspirated under continuous rinsing. In our experience it is very important to vaporize the remaining cyst capsula with radiofrequency or laser application. The success of the intervention is controlled intraoperatively by digital dynamic subtraction cystography combined with ventriculography. In patients with cystic glioma, where a resection of the solid tumour parts is planned, an Ommaya reservoir for repeated fluid aspiration is installed.

Results

From June 1990 to August 1994 more than 70 patients with intracerebral cystic space occupying lesions were operated on in stereotactic endoscopic technique. There was no operative mortality, operative morbidity was 1.4%. Malignant cystic neoplasms (anaplastic astrocytomas, glioblastomas, metastases) were operated on to establish histopathologic diagnosis, to reduce raised ICP, and to apply reservoir systems for repeated puncture. In some cases stereotactic endoscopic intervention was followed by microsurgical resection of the solid tumour parts and radiotherapy. Stereotactic endoscopic operations for benign intracerebral space occupying lesions were performed with the aim of a definite cure. Intraventricular cysts were treated most successfully.

a) Colloid Cyst

Seven patients with colloid cysts were operated on in stereotactic endoscopic technique. Three patients had symptoms according to group 1 (headache, papilledema, and neurologic disorders) and four patients according to group 3 (headache, drop attacks) defined by Kelly [14]. Total evacuation and vaporization of the capsule was performed in all cases. There was one patient with an intraoperative hemorrhage from the tumour capsule. Six months after the intervention this patient is suffering from memory deficits and slight loss of impulsion. Up to now there was no cyst recurrence

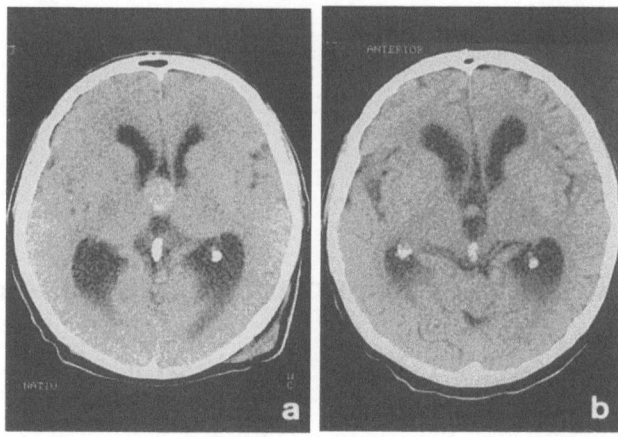

Fig. 1. (a) CCT: colloid cyst of the third ventricle. (b) After stereotactic endoscopic intervention CT control shows only a small rest of the cyst capsule

(Fig. 1a, b). In four more patients, who were admitted to our department with the radiological diagnosis "colloid cyst", stereotactic endoscopic biopsy of the intraventricular tumour revealed two ependymomas, one plexuspapilloma, and one solid craniopharyngioma.

b) Cystic Craniopharyngioma

In five patients with cystic craniopharyngioma, the large space occupying intraventricular tumour part was evacuated through a precoronar burrhole approach. The cyst wall was shrinked by laser beam or radiofrequency as much as possible. If parts of the capsule were adherent to parasellar structures they were left in situ. 3 patients are now without tumour recurrence, whereas in two patients nine and twelve months after the endoscopic intervention the remaining solid tumour portion was removed in microsurgical technique (Fig. 2a, d).

c) Pineal Cyst

Cystic lesions of the pineal gland caused periodical headache and vertigo in two patients. Neuroradiological examination showed the large space occupying cystic lesion with a hydrocephalus due to an aqueduct stenosis. The cysts were punctured and the contents were aspirated. Adherent parts of the capsule were left in situ. The restoration of normal CSF circulation was controlled intraoperatively by dynamic digital subtraction ventriculography. Both patients are without neurological deficits or cyst recurrence (Fig. 3a, b).

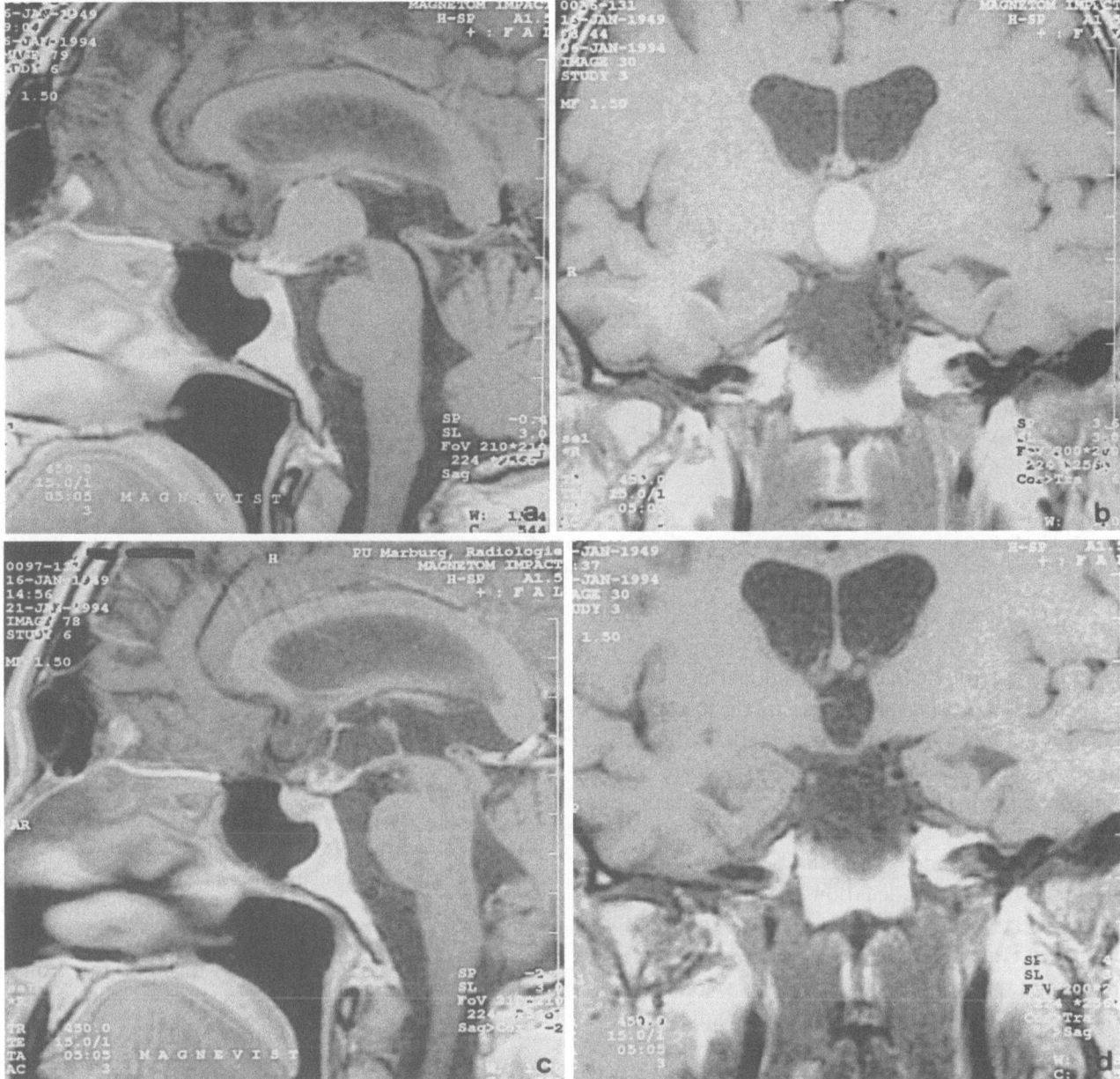

Fig. 2. (a, b) Sagittal and coronar MRI: suprasellar cystic lesion in the anterior part of the third ventricle growing up from the suprasellar region. (c, d) After evacuation of the cystic craniopharyngioma control MRI shows the residual tumour capsula

Discussion

Stereotactic endoscopic interventions for cystic malignant intracerebral space occupying lesions are performed to guarantee a high rate of acute decompression using a minimally invasive approach and at the same time to establish histopathological diagnosis. The main goal for the intervention is to save patient's quality of life for the remaining short space of life. The decision for a secondary microsurgical tumour-resection should be made with regard to the patient's general state. This therapeutic strategy is evident. In contrast, the treatment of benign intracerebral/intraventricular cystic lesions appears more controversial. In our opinion intraventricular cysts are particularly omitable for stereotactic endoscopic intervention using a flexible steerable endoscope together with the Marburg Neuroendoscope Fixation and Guiding System.

Colloid Cysts

Since the first extirpation of a colloid cyst performed by Dandy in 1921 various operative techniques in the

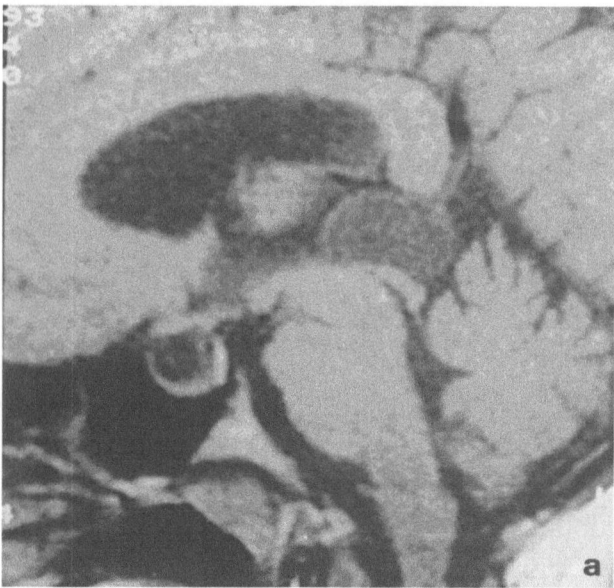 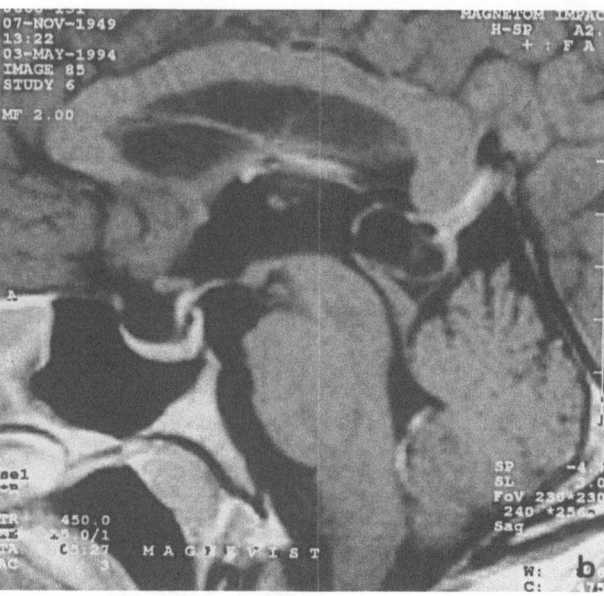

Fig. 3. (a) MRI: a large space occupying lesion in the pineal region which causes a compression of the proximal aqueduct of Silvii. (b) MRI control after stereotactic endoscopic cyst evacuation with the remaining tumour capsule. Clearly visible the decompressed aqueduct

therapy of this pathology have been proposed [1,2,6,15,18]. Before the microsurgical era operative mortality was 10–20% [8]. Conventional microsurgical approaches to resect colloid cysts are the frontal transcortical method [20], which uses the lateral ventricle as a corridor, and the interhemispheric transcallosal approach [25]. The complication rates of both techniques amount to 5% (the number of CSF diversion system is not included). In 1978 Bosch reported about the first successful evacuation of colloid cysts using stereotactic techniques [6]. He was followed by Donauer, Konziolka, Ostertag, and Warnke, who have the largest series of colloid cysts treated stereotactically partly combined with endoscopic technique [8,15, 18, 26]. The reported complication rates of stereotactic colloid cyst treatment are extremely low, though many of the patients were treated preoperatively with CSF shunting systems. Compared to stereotactic operative methods our stereotactic endoscopic evacuation technique of colloid cysts has the advantage of direct visual control to prevent injury of the thalamostriate and septal veins and the choroid plexus, which can cover the upper part of the cyst [4,5]. Furthermore the complete removal of the cyst contents can be judged intraoperatively and the remaining cyst wall is vaporized stepwise under endoscopic guidance. The main point of discussion is that remaining parts of the cyst wall can lead to a recurrence [17]. In our experience and in long term follow-up studies of patients with colloid cysts, which were operated on with stereotactic technique, this assumption could not be verified [18,26].

Cystic Craniopharyngioma

Despite the use of microsurgical techniques mortality and morbidity in operative treatment of craniopharyngioma are not negligible. Operative mortality after primary resection is about 5% [16,21,23], whereas after a second intervention mortality increases to 30% [24]. The main problem in craniopharyngioma surgery is recurrence. Even in patients who were operated on radically, recurrence rate is high [24]. As a consequence we propose to operate on cystic craniopharyngioma using minimal invasive approaches. The tumour is reached easily through the frontal precoronar burrhole approach without major tissue traumatization. It is punctured and evacuated to guarantee a sufficient inner decompression of hypothalamus, optic nerve and other surrounding, sensitive structures. The cyst wall and additional microcystic changes are vaporized. In case of tumour recurrence microsurgical resection, irradiation, and stereotactic application of a cysto-ventricular shunt or Ommaya reservoir are recommended [10,11].

Pineal Cyst

Symptomatic pineal cysts are rare [9]. The main indication for operative treatment of space occupying pineal cysts is the development of an acute occlusive hydrocephalus. Microsurgical approaches as the infratentorial, supracerebellar route to the pineal gland have a mortality of about 5% and a morbidity of 8%

[22]. In 1994 Fain reported on 24 patients suffering from symptomatic pinealis cyst with an operative morbidity of 12% [9]. With the stereotactic endoscopic evacuation technique, mortality and morbidity are reduced to zero. The main disadvantage is, as it is the same with colloid cysts and cystic craniopharyngioma, that adherent parts of the cyst wall cannot be resected using this approach. However, up to now there is no cyst recurrence in the two patients with pineal cysts we have operated on.

Conclusion

Stereotactic endoscopic interventions for cystic intracerebral lesions are minimal invasive and have a lower complication rate than microsurgical operations. The main advantage to ordinary stereotactic procedures is direct visual control of the intervention enabling an intraoperative control of the cyst evacuation and resection or vaporization of the cyst wall. With intraoperative dynamic digital subtraction ventriculography intraoperative restoration of CSF pathways can be demonstrated.

References

1. Abernathey CD, Davis DH, Kelly PJ (1989) Treatment of colloid cysts of the third ventricle by stereotaxic microsurgical laser craniotomy. J Neurosurg 70: 525–529
2. Abtunes JL, Louis KM, Ganti SR (1980) Colloid cysts of the third ventricle. Neurosurgery 7: 450–455
3. Bauer BL, Hellwig D (1992) Minimally invasive neurosurgery I. Acta Neurochir (Wien) [Suppl] 54
4. Bauer BL, Hellwig D (1994) Intracerebral and intraspinal endoscopy. In: Schmidek HH, Sweet WH (eds) Operative neurosurgical techniques, 3rd Ed. Saunders, Philadelphia.
5. Bauer BL, Hellwig D, Sweet WH, Schmidek HH (1994) The management of intracranial arachnoid, suprasellar and rathke's cleft cysts. In: Sweet WH, Schmidek HH (eds) Operative neurosurgical techniques, 3rd Ed. Saunders, Philadelphia
6. Bosch DA, Rahn T, Backlund EO (1978) Treatment of colloid cysts of the third ventricle by stereotactic aspiration. Surg Neurol 9: 15–18
7. Bruce JN, Stein BM (1992) Infratentorial approach to pinealis tumours. In: Wilson CB (ed) Personal approaches to classic operations. Williams and Wilkins, Baltimore, pp 63–75
8. Donauer E, Moringlane JR, Ostertag CB (1986) Colloid cysts of the third ventricle. Open operative approach or stereotactic? Acta Neurochir (Wien) 83: 24–30
9. Fain JS, Tomlinson FH, Scheithauer BW, et al (1994) Symptomatic glial cysts of the pineal gland. J Neurosurg 80: 454–460
10. Fox JL (1967) Intermittent drainage of intracranial cysts via the subcutaneous Ommaya reservoir. Technical note. J Neurosurg 38: 251–256
11. Gutin PH, Klemme WM, Lagger RL, et al (1980) Management of unresectable cystic craniophayngeoma by aspiration through an Ommaya rerservoir drainage system. J Neurosurg 52: 36–40
12. Hellwig D, Bauer BL, List-Hellwig E, et al (1991) Stereotactic-endoscopic procedures on processes of the cranial midline. Acta Neurochir (Wien) [Suppl] 53: 23–32
13. Hellwig D, Bauer BL (1992) Minimally invasive neurosurgery by means of ultrathin endoscopes. Acta Neurochir (Wien) [Suppl] 54: 63–68
14. Kelly R (1951) Colloid cysts of the third ventricle. Analysis of twenty-nine cases. Brain 74: 23–65
15. Kondziolka D, Lunsford LD (1991) Stereotactic management of colloid cysts: factors predicting success. J Neurosurg 75: 45–51
16. Laws ER (1986) Craniopharyngeoma. Neurosurgery 19: 326
17. Mathiesen T, Grane P, Lindquist C, et al (1993) High recurrence rate following aspiration of colloid cysts in the third ventricle. J Neurosurg 78: 748–752
18. Ostertag CB (1990) Surgical techniques in the management of colloid cysts of the third ventricle – the stereotactic endoscopic approach. In: Symon L (ed) Advances and technical standards in neurosurgery, Vol 17. Springer, Wien New York, pp 143–149
19. Raimondi AJ (1987) Pediatric neurosurgery. Theoretic principles. Art of surgical techniques. Springer Berlin, Heidelberg New York Tokyo, p. 229
20. Shucart WA, Stein BM (1978) Transcallosal approach to the anterior ventricular system. Neurosurgery 3: 339–343
21. Sung OP, Chung CH, Harisiadis L, et al (1981) Treatment results of craniopharyngeomas. Cancer 47: 852–874
22. Stein BM (1971) The infratentorial supracerebellar approach to pineal lesions. J Neurosurg 35: 197–202
23. Stein WH (1976) Radical treatment of craniopharyngeomas: therapeutic alternatives. Clin Neurosurg 23: 52–79
24. Sweet WH (1980) Recurrent craniopharyngeomas: therapeutic alternatives. Clin Neurosurg 27: 206–229
25. Symon L, Pell M (1990) Surgical techniques in management of colloid cysts of the third ventricle. The transcortical approach. In: Symon L (ed) Advances and technical standards in neurosurgery, Vol 17. Springer, Wien New York, pp 122–133
26. Warnke PC, Hans FJ, Jaiswa V, et al (1994) Stereotactic endoscopic therapy of colloid cysts. In: Bauer BL, Klinger M, Brock M (eds) Advances in neurosurgery, Vol 22. Springer, Berlin Heidelberg New York Tokyo, pp 134–139

Correspondence: D. Hellwig, M.D., Department of Neurosurgery, Philipps University Marburg, D-35033 Marburg, Federal Republic of Germany.

Acta Neurochir (1995) [Suppl] 64: 64–68

Neurosurgery for Affective Disorders at Atkinson Morley's Hospital 1948–1994

N. Kitchen

Department of Neurosurgery, Atkinson Morley's Hospital, Copse Hill, Wimbledon, London, U.K.

Summary

Neurosurgical practice in the treatment of affective disorders has changed dramatically over the last 40 years. This paper traces the changes which have occurred in one institution, namely Atkinson Morley's Hospital, Wimbledon, London, UK from 1948 to the present day.

Freehand operations designed by McKissock and performed on large numbers of patients disappeared as better non-surgical treatments became available and long term complications and treament failures became dearer. In the 1970's stereotactic limbic leucotomy, a much more focal and accurate operation, was devised and became popular. The present day practice utilises precisely the same techniques as the original stereotactic limbic leucotomy but is employed on small numbers of patients. The procedure continues to have a role for those few patients with severe psychiatric illness, particularly obsessive compulsive disorder, which has proved refractory to other therapeutic modalities.

Keywords: Stereotaxy; psychosurgery; obsessive-compulsive-disorder.

Introduction

Neurosurgery for affective (i.e. psychiatric) disorders in the UK has changed almost beyond recognition over the last 40 years in a large part due to medical advances (such as the introduction of effective drug and behavioural treatments), but importantly, these changes have taken place on the foreground of a changing socio-political milieu. This paper aims to describe such changes by tracing the neurosurgical treatment of affective disorders in one institution, namely Atkinson Morley's Hospital, Wimbledon, London, UK.

There have been essentially three distinct phases represented by:

1. Sir Wylie McKissock – 1950s and 1960s.
2. Stereotactic limbic leucotomy – 1970s.
3. Present day practice.

1. Sir Wylie McKissock (1906–1994) (Fig. 1)

The late Sir Wylie McKissock was the leading British neurosurgeon of his generation influencing neurosurgical practice and organisation in the UK to a considerable degree. He will be remembered for many different things by many different people.

Sir Wylie was a strong proponent of neurosurgery for affective disorders. His practice reflected that which was common during the 1950s in the UK. Thus, surgery was performed mostly in local psychiatric hospitals and not in the neurosurgical centre. This was partly because most of the patients were chronically institutionslised but also, more probably, because the operative procedures themselves were very simple to perform, required no special instrumentation and the patients were perceived to require no special post-operative neurosurgical or nursing care.

After performing the first rostral leucotomy as an open procedure in December 1948 McKissock quickly changed to a "closed" procedure (i.e. performed via bilateral burr holes). This procedure was similar to Scoville's operation which involved undercutting of the superior frontal convolution though McKissock thought his was less in extent (the leucotomy passed from just in front of the coronal suture to the frontal pole and measuring 1–2 cm in width). However, Sir Wylie emphasised the "experimental nature of leucotomy" [6] and tried several different angles of cutting of the frontal white matter ranging from the original Rostral Leucotomy to the Rostral G (after Grantham who was then placing electrocoagulative lesions in much the same place. The place of the Rostral G passed to the junction of the middle and posterior thirds of the orbital plate) and finally to the Rostral P (after Pippard, one of his referring psychiatrists. The cut in this

Fig. 1. Sir Wylie McKissock (1906–1994). This portrait, commisioned around the time of his retirement, currently hangs in the Wolfson Centre, Atkinson Morley's Hospital

operation was further forward than the Rostral G but behind the original Rostral Leucotomy and aimed for the midpoint of the orbital roof). The Rostral P leucotomy became McKissock's favoured procedure as he thought it combined lesser side-effects without compromising efficacy as compared with the earlier operations he had used. At a meeting of the Royal Society of Medicine in 1956 [6] Sir Wylie described having personally performed 490 such operations for patients with various psychiatric complaints (and it is certain that many more were performed by him by the time of his eventual retirement over a decade later). In this 1959 communication 46.8% of all patients were reported to have been able to be discharged and return to work, 19.3% discharged but not able to work and 29.8% were still in-patients. Best results were achieved in depressives with obsessional components, the worst in chronic schizophrenics. The simple (some would say rudimentary) operative procedures were safe with a low neurological morbidity and with an operative mortality of 1%. However, the incidence of epilepsy at long term follow-up was high (probably over 20%) and rigorous long term follow-up in terms of the patients' cognitive and emotional status was lacking and it is certain that many patients were left considerably impaired in these respects. Also no objective measure-

ments of efficacy in terms of the management of the psychiatric condition was made except in the crudest way. What these destructive operations probably did achieve was to make patients more manageable at a time effective alternative treatments were not available. A decade or so later drug treatments had largely taken over.

2. Stereotactic Limbic Leucotomy [4, 5, 7]

The operation of stereotactic limbic leucotomy was devised at Atkinson Morley's Hospital by the neurosurgeon Mr A. E. Richardson and his collaborative psychiatrist Dr D. Kelly in the early 1970s. It was, and remains, unique in the field of surgery for affective disorders in that it combined lesions in the cingulum bundle (derived from Fulton's experimental work and Cairn's clinical use [12]. Pure cingulotomy is still practised in some centres [2]) with those in the frontal lobes. Lesions were placed using traditional stereotactic techniques to produce exactly reproducible focal lesions at these two areas in order to disrupt the reverberating limbic circuits (Papez's loop [9]) and the fronto-limbic and thalamocortical connection [1, 8].

Four lesions were made in the anterior cingulum bundle bilaterally and 3 lesions in the ventromedial quadrant of the frontal lobes bilaterally (thus making 14 in all). Using a thermistor electrode with a 8×1 mm tip heated to 56 centigrade for 2 minutes at each of these 14 sites the effect aws essentially to produce 1 large lesion at each of 4 sites (i.e. cingulum bundle and inferior-medial frontal lobe white matter bilaterally).

All the lesions were placed via 2 parasaggital coronal burr holes. General anaesthesia was employed and air ventriculography combined with biplanar teleradiology used to calculate the target sites (a combination of boney and ventricular landmarks were used. Figs. 2a and b).

Electrophysiological corroboration was employed. The patient was allowed to breath spontaneously. Stimulation of the target sites produced a number of autonomic changes (Fig. 3), most notably arrest of respiration.

Results were rigorously and scientifically scrutinised on all patients up to at least 20 months post-operatively using a number of creditable psychometric testscoring systems (Tables 1 and 2). On all the numerous outcome measures the results were impressive with over two-thirds of all patients improved (Table 1). Results were best in patients with obsessive-compulsive disorders or in those with unipolar depressive

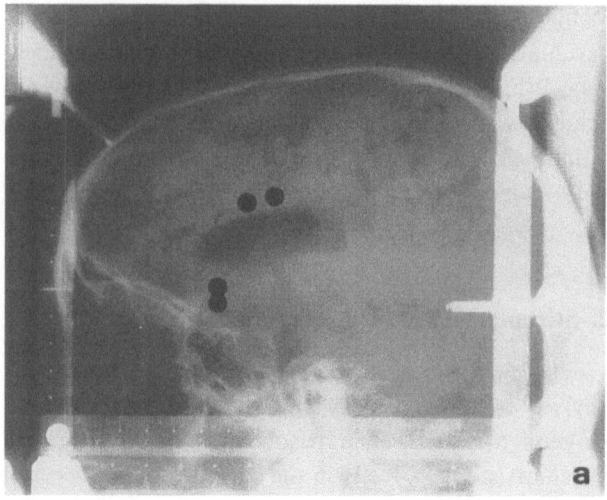 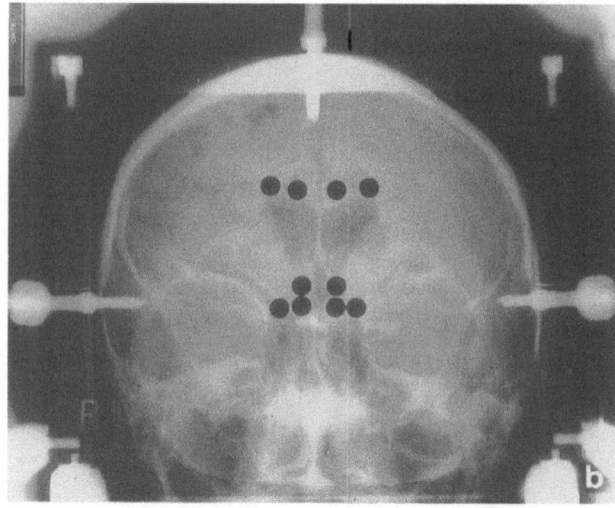

Fig. 2 (a, b). Stereotactic air ventriculograms. (a) AP, (b) lateral. A Leksell frame is used with biplanar teleradiology. Air is introduced into the ventricular system via a sitting lumbar puncture. The targets have been calculated and are each represented by a black circle (approximate size of lesion).

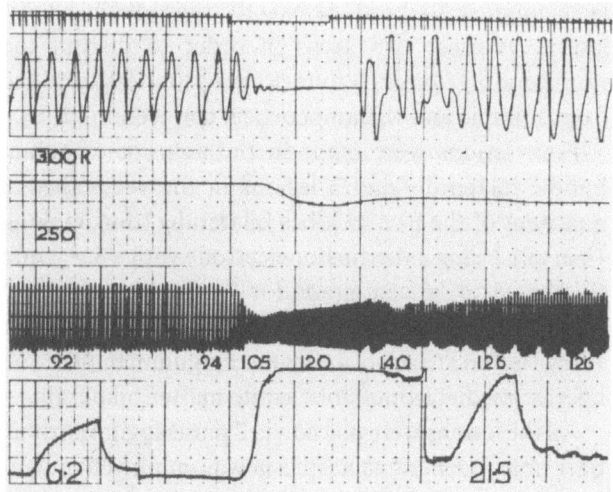

Fig. 3. Autonomic changes seen on stimulation of the ventro-medial frontal targets during limbic leucotomy (similar changes are found at the cingulum target sites). Channel 1: time (secs) depressed during stimulation. Channel 2: respiration; apnoea followed by tachypnoea. Channel 3: Skin resistance; falls, Channel 4: finger pulse; tachycardia associated with a decreased pulse pressure during stimulation. Channel 5: Forearm blood flow. (Taken from Anxiety and Emotions, Kelly D[4], p235. With grateful permission)

illnesses with obsession features. At least 150 patients were thus treated by the mid 1970s. Though prospective these studies were not blinded and independant observers were not used. As referral patterns changed and behavioural therapies became more popular on the one hand and interest in functional neurosurgery dwindled on the other the numbers of patients under-

going surgery for affective disorders declined dramatically.

3. Current Surgical Practice

A small but relatively constant number of patients with affective disorders are currently referred to Atkinson Morley's Hospital for neurosurgery. Since 1987 24 patients with various psychiatric diagnoses have undergone stereotactic limbic leucotomy; 17 with obsessive-compulsive disorder, 5 with endogenous depression, and 2 suffering from Gilles de la Tourette Syndrome. Two patients with obsessive compulsive disorder were referred following failed cingulotomies elsewhere.

Following stereotactic limbic leucotomy the mean in-patient stay at Atkinson Morley's Hospital is currently 4 days after which they are transfered to a local psychiatric hospital for a short period of behavioural therapy and rehabilitation.

With a mean follow-up of 35 months the overall outcome has been as follows; 13 have been much improved or are symptom free, 5 have been improved to some extent whilst 6 are no better or are worse. 1 patient with endogenous depression commited suicide 3 nonths following surgery. The best results have been experienced in those patients with obsessive compulsive disorder. These results broadly duplicate the findings of the 1970's cohort of patients undergoing the identical procedure. In addition we have found very good results in our two patients with Gilles de la

Table 1. *Clinical Ratings 6 Weeks and 20 Months Following Stereotactic Limbic Leucotomy.* (Taken from Anxiety and Emotions, Kelly D[4], p245. With grateful permission)

	6 Weeks						20 Months					
	N	I	II	III	I V	V	% of total improved	I	II	III	I V	V	% of total improved
Obsessional neurosis	49	2	22	14	11	0	78	7	23	11	5	3	84
Chronic anxiety	27	0	10	8	8	1	67	3	5	9	8	2	63
Depression	36	5	12	9	10	0	72	7	7	8	11	3	61
Schizophrenia	19	0	4	7	7	1	58	0	6	6	5	2	63
Other:												
Personality disorder	7	0	0	1	6	0		0	3	2	2	0	
Anorexia nervosa	5	0	0	3	2	0		1	0	0	3	1	
Depersonalization	1	0	0	0	0	1		0	0	0	0	1	
Intractable pain	1	0	1	0	0	0		0	0	1	0	0	
Dementia (depression)	1	0	0	0	1	0		0	0	0	0	1	
Paranoia	1	0	0	1	0	0		0	0	0	1	0	
Palilalia (Parkinsonism)	1	0	0	0	1	0	— 35	0	0	0	1	0	— 41
Total	148	7	49	43	46	3	67	18	44	37	36	13	67

I symptom free, *II* much improved, *III* improved, *IV* not improved, *V* worse.

Table 2. *Psychometric Values Before and 20 Months After Stereotactic Limbic Leucotomy.* (Taken from Anxiety and Emotions, Kelly D[4], p251. With grateful permission)

		20 Months mean		
	Pre-	Post	Difference	p[a]
Anxiety				
Taylor	32.3	23.7	8.6	0.001
Hamilton	26.3	16.4	9.9	0.001
MHQ anxiety	11.6	8.6	3.0	0.001
Phobic	7.5	5.9	1.6	0.001
Somatic	8.6	5.7	2.9	0.001
Basal self rating	5.3	4.5	0.8	0.01
observer rating	5.3	4.0	1.3	0.001
Stress self rating	6.9	6.5	0.4	NS[b]
observer rating	6.7	6.0	0.7	0.001
Depression				
Beck	28.0	18.3	9.7	0.001
Hamilton	25.1	15.2	9.9	0.001
HAQ	10.3	7.7	2.6	0.001
self rating	6.3	4.4	1.9	0.001
Observer rating	6.2	3.8	2.4	0.001
Obsessions				
Symptoms	26.5	19.3	7.3	0.001
Traits	12.1	9.5	2.6	0.001
Resistance	35.4	22.4	13.0	0.001
Interference	41.1	22.2	18.9	0.001
MHQ	10.8	8.7	2.2	0.001
Neuroticism	•			
MPI	34.0	26.8	7.2	0.001
Psychiatric symptoms				
Cornell	24.9	16.2	8.7	0.001
Extroversion				
MPI	13.3	16.0	+ 2.7	0.01
Hysterical				
MHQ	5.9	4.8	1.1	0.001

[a] *p*: correlated groups *t* test.
[b] *NS*: not significant.

Tourette Syndrome [10, 11]. Complications have been few. Five patients suffered transient urinary incontinence which resolved by the time of discharge, 4 patients experienced temporary lethargy and 1 was frankly confused. Finally 1 patient suffered a facial palsy which had resolved by 6 months following surgery. We have found follow-up MRI extremely useful in further delineating the target site anatomy in individual cases (Fig. 4a–c).

In small numbers of patients with severe and life threatening or incapacitating obsessive-compulsive disorder stereotactic limbic leucotomy remains a reasonable treatment option and can be effective when all other therapies hace failed.

Conclusion

Neurosurgery is the most conservative of all the surgical disciplines and no more so than in the sensitive area of neurosurgery for affective disorders. Over the last 40 years the surgical practice in this field at Atkinson Morley's Hospital has changed dramatically; the freehand operations performed on large numbers of patients disappeared as better non-surgical treatments became available and long term complications and treament failures became clearer. Stereotactic limbic leucotomy designed over 20 yers ago has proven effcacy and safety. As a result the present day practice utilises precisely the same techniques as the originaloperation but is employed on far fewer patients.

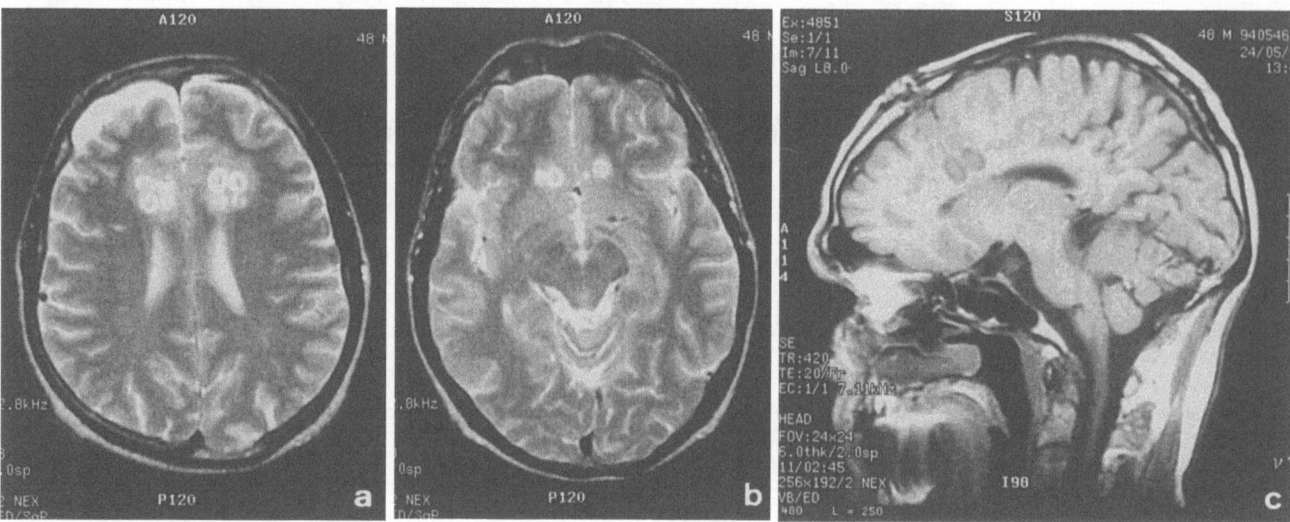

Fig. 4 (a–c). Post-operative MRI studies. The T2 axial images show the cingulate (a) and the frontal (b) lesions separately. The T1 saggital image show both sets of lesions (the cingulate lesions more distinct in this image, though frontal changes can be seen) and the burr hole and electrode track

The problems facing surgery for affective disorders in the future are clear; any innovation is morally prohibited if seen as "experimentation". Yet the only way to definitively test the efficacy of such surgery is to conduct prospective properly blinded trials. However, the obstacles are great with very small numbers of operations being performed at very few centres using operations often unique to the institution [3]. At present in the UK small numbers of incapacitated and truly refractory patients are undergoing neurosurgery for affective disorders (at Atkinson Morley's Hospital the number has remained at about 4 per year). At present the neurosurgeon acts as the technician; the Mental Health Commissioners and independant psychiatrists decide on the suitability of referral and the ability of the patient give consent. Yet it is essential if further improvements are to be made in the surgical treatment of affective disorders for clinical and research collaborations between interested surgeons and psychiatrists continue and flourish.

References

1. Alexander GE, Crutcher MD, Delong MR (1990) Basal ganglia-thalamocortical circuits: parallel substrates for motor, oculomotor, "prefrontal" and "limbic" functions. Prog Brain Res 85: 119–146

2. Ballantine HT, Boukoms AJ, Thomas EK, Giriunas IE (1987) Treatment of psychiatric illness by stereotactic cingulotomy. Biol Psychiatr 22: 807–809

3. Bridges PK, Bartlett JR (1977) Psychosurgery: yesterday and today. Br J Psychiatr 131: 249–260

4. Kelly D (1980) Anxiety and emotions. Thomas, Springfield, Illinois

5. Kelly D, Richardson A, Mitchell-Heggs N (1973) Stereotactic limbic leucotomy: neurophysiological aspects and operative technique. Br J Psychiatr 123: 133–140

6. McKissock W (1959) Discussions on psychosurgery. Proc Roy Soc Med 52 206–209

7. Mitchell-Heggs N, Kelly D, Richardson AE (1976) Stereotactic limbic leucotomy – a follow-up at 16 months. Br J Psychiatr 128: 226–240

8. Modell JG, Mountz JM, Cutis GC, Greden JF (1989) Neurophysiologic dysfunction in basal ganglia/limbic striatal and thalamocortical circuits as a pathogenetic mechanism of obsessive-compulsive disorder. J Neuropsychiatr 1: 27–36

9. Papez JW (1937) A proposed methanism of emotion. Arch Neurol Psychiatr 38: 725–743

10. Robertson M, Doran M, Trimble M, Less AJ (1990) The treatment of Gilles de la Tourette syndrome by limbic leucotomy. J Neurol Neurosurg Psychiatry 53: 691–694

11. Sawle GV, Less AJ, Hymas NF, Brooks DJ, Frackowiak RSJ (1993) The metabolic effects of limbic leucotomy in Gilles de la Tourette syndrome. J Neurol Neurosurg Psychiatry 56: 1016–1019

12. Whitty CWM, Duffield JE, Tow PM, Cairns H (1952) Anterior cingulectomy in the treatment of mental disease. Lancet i: 475–481

Correspondence: Neil Kitchen, FRCS, Department of Neurosurgery, Atkinson Morley's Hospital, Copse Hill, Wimbledon, London, U.K.

Acta Neurochir (1995) [Suppl] 64: 69–73

Frontal Cingulotomy Reconsidered from a WGA-HRP and c-Fos Study in Cat

R. Kuroda, A. Yorimae, Y. Yamada, Y. Furuta, and **A. Kim**

Department of Neurosurgery, Kinki University, Osaka-Sayama, Japan

Summary

A recent positron emission tomography (PET) study demonstrated that the anterior cingulate cortex (area 24), in addition to SI and SII cortices, was activated by painful stimuli. In order to elucidate the participation of relay nuclei in the ascending pain pathway to area 24, we performed a regrograde labelling study with WGA-HRP injection into area 24 in cats. Area 24 was found to receive pain-related thalamic inputs from the intralaminar nuclei including the central medial nucleus, midline nuclei, modiodorsal nucleus and possibly the submedial nucleus. We then examined the expression of Fos protein in CNS induced by formalin injection into the face in cats. Fos positive neurons were demonstrated in areas 23 and 24, the anterior limbic area, insular cortex, midline and paraventricular nuclei in the thalamus, paraventricular nucleus and other areas in the hypothalamus, and in many nuclei in the brainstem in both the formalin-injected group and the control group (anesthesia only). Labelled regions appeared to correspond to stress-related sites. The sole difference from the control group was the expression of Fos in the coronal gyrus and in the trigeminal caudalis nucleus in the experimental group. Although more Fos positive cells were observed in area 24 in experimental than in control cats, the difference was not significant. Our findings suggest that the demonstrated response of area 24 on PET scan represents stress- and emotion-related events rather than pain. Surgical intervention into the anterior cingulate cortex including cingulotomy thus appears to relieve stress and emotion associated with chronic pain, but not pain inself.

Keywords: Pain; area 24; WGA-HRP; c-fos; cingulotomy.

Introduction

A recent functional study in man by positron emission tomography (PET) demonstrated that the anterior cingulate cortex (area 24) along with SI and SII cortices may participate in the perception of pain [11]. However, there is in the literature different opinions regarding the fiber connections between area 24 and thalamic terminal sites of the spinothalamic tract [9, 14]. In recent years it has been demonstrated that expression of the proto-oncogene c-fos (Fos protein) in CNS cells may serve as a "third messenger" in long-term responses to pain stimuli [3] as well as a metabolic marker of brain activity [2]. In this study we used histochemical and immunohistochemical methods for the examination of 1) retrograde labelling of neurons in the thalamus after injection of wheat germ agglutinin conjugated to horseradish peroxidase (WGA-HRP) into area 24, and 2) c-fos expression in the CNS, particularly in area 24, after injection of formalin in the face of cats. On the basis of our findings, we discuss frontal cingulotomy, originally proposed by Foltz and White [4] based on the "circuit of emotion" theory of Papez [10].

Materials and Methods

1) Retrograde Labelling WGA-HRP

In four cats a small dose of WGA-HRP was stereotactically injected into area 24 of the cingulate gyrus from the contralateral hemisphere in order to avoid leakage of dye into other regions of cortex. After a 48-hour survival period, the animals were perfused and fixed with 8% formalin in 0.1 M phosphate buffer under deep anesthesia. Serial 50 μm frozen sections of the brain were processed by the tetramethylbenzidine method [8]. Half of the sections were counterstained with neutral red for identification of the cytoarchitecture. The location of area 24 was determined according to the criteria supplied by Rose and Woolsey [12].

2) Immunohistochemical Study of Expression of Fos Protein

After induction of urethane anesthesia by intraperitoneal injection, a small dose of 5% formalin was injected subcutaneously as a strong pain stimulus into the right ophthalmic region in two cats and into the right maxillary region in another two cats. No treatment except anesthesia was performed on two other cats serving as controls. After a two-hour survival period, all six animals were perfused and fixed with 4% paraformaldehyde in 0.1 M phosphate buffer. The brains were removed and fixed with the same fixative overnight. After immersion in 30% sucrose solution, serial 30 μm sections were cut. Fos protein was demonstrated immunohistochemically by a peroxidase/anti-peroxidase (PAP) method using c-fos antibody at 1:500 dilution as primary antibody (Oncogence Science, Ab-2), with reaction for 72 hours at 4°C. Fos protein was subsequently visualized by diaminobenzidine reaction.

Results

1) Thalamic Afferents to Area 24 (Table 1)

WGA-HRP injection into area 24 was confirmed in all four cats, as shown in the diagrams at the top of Table 1. A small amount of diffusion of the dye into area 23 in the caudal portion of the cruciate gyrus was observed in cases 225 and 312.

The cells of the anterior nuclear group were all strongly labelled in all four cats. The rhomboid nucleus (RH) and the ventromedial nucleus (MV) of the midline group were also moderately labelled. In the intralaminar nuclear group, many cells of the central medial nucleus (CMN) (Fig. 3A) and paracentral (PAC) uncleus were consistently labelled. Labelling of the lateral central nucleus (CL), centromedian nucleus (CM) and parafascicular nucleus (Pf) was inconsistent. Labelling of a few cells was observed in all cases in the lateral portion of the mediodorsal nucleus (MD). In the ventral nuclear group, no labelled cells were detected in the posterolateral ventral nucleus (VPL). Interestingly, a few cells in the submedial nuclus (SUM) were labelled in two of the four cats. Labelling was also consistently present in the anterior ventral nucleus (VA), lateral posterior nucleus (LP) and lateral dorsal nucleus (LD).

Table 1. *Distribution of Labelled Cells in the Thalamus Following Injection of Retrograde Tracer in Area 24*

Anterior nuclei				
AM	+ +	+	+ +	+
AD	+	+ +	+	±
AV	+ +	+ +	+ +	±
Midline nuclei				
RH	+	+	+	±
MV	±	±	+	−
Intralaminar nuclei				
CMN	+ +	+	+ +	+
CL	+	−	±	±
PAC	+	±	+	+
CM	±	±	+	−
Pf	±	−	±	+
Medial nuclei				
MD	±	±	±	±
Ventral nuclei				
VA	+	±	+	±
VL	±	−	−	−
VPL	−	−	−	−
VMP	±	−	−	−
SUM	±	−	±	−
Lateral nuclei				
LD	+ +	+	+	±
LP	+ +	±	+	±

Figures on top of the table show injection sites in 4 cats. Each arrow points to injection site (black area) from the contralateral hemisphere. The surrounding stippled area shows slight diffusion area of the tracer. Symbols: + +: 10 or more cells in each section; +: less than 10 cells in each section; ±: a few cells in several sections; −: absent. For abbreviations, see text.

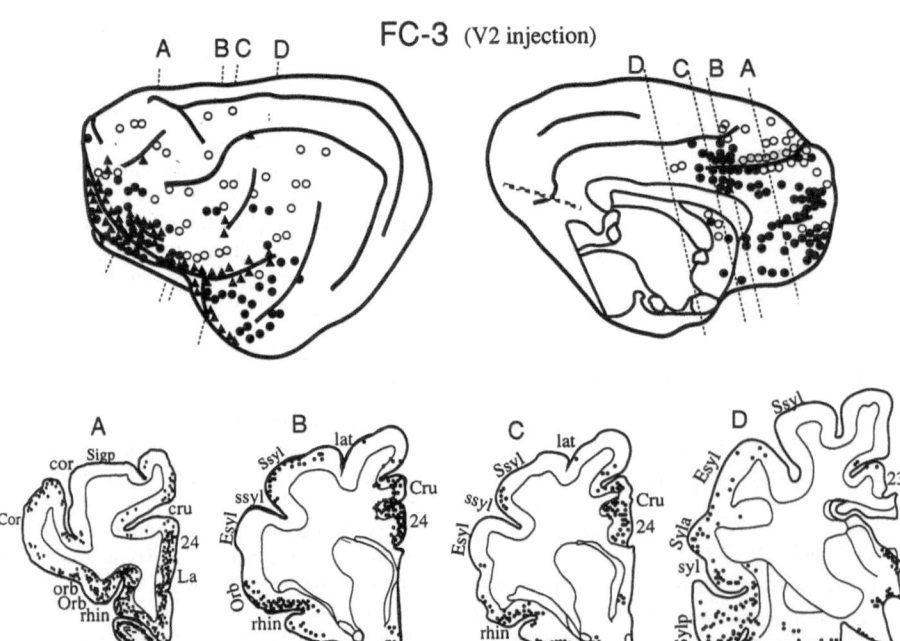

Fig. 1. Cortical distribution of Fos positive neurons in cat FC-3 which had formalin injected into the maxillary region. Upper figures shows distribution of Fos in reconstructed cortex. For symbols, see Fig. 2. Lower figures show Fos positive cells in representative sections. Each dot represents a single stained neuron. For abbreviations, see text

2) Distribution of Fos Positive Neurons in the CNS

The cortical distribution of Fos positive cells in a representative case, FC-3, of the experimental group, which had formalin injected into the mixillary region, is shown in Fig. 1. Positive neurons were densely distributed in the proreal gyrus which corresponds to prefrontal cortex in man, and in piriform cortex (Pir), orbital gyrus (Orb), rhinal sulcus (rhin) and sylvian gyri (Syla, Sylp). Each of the latter gyri and sulci is part of the olfactory and insular cortex [6]. In the coronal gyrus, where facial sesation is represented, a moderate number of Fos positive neurons (Fig. 3C) were detected anteriorly. A moderate number of labelled cells were also found in the posterior portion of the anterior ectosylvian gyrus (Esyl) and the superior portion of the anterior gyrus (Syla), which correspond to auditory cortex (AI, AII). On the medial surface of the cortex, many immunoreactive neurons were observed in the anterior cingulate cortex (area 24) (Cin, 24) (Fig. 3B), as well as in the anterior limbic cortex (La). Labelled cells were also densely distributed in the posterior portion of the cruciate gyrus (Cru), which has been designated area 23 by some authors [12], but area 3a by others [5]. In the anterior portion of area 23 in the parasplenial gyrus (Paraspl), labelled cells were present as dencely as in area 24, whereas few cells were found in the posterior portion. In the other three cats of the experimental group, the cortical distribution of Fos was more or less the same as in FC-3. However, Fos cells in cats which had formalin injected into the ophthalmic region were located more posteriorly in the coronal gyrus than those in case FC-3. In the control group, the distribution of Fos positive cells was almost the same as that in the experimental group, except for the absence of Fos cells in the coronal gyrus. Figure 2 shows the cortical distribution in a representative case. Statistical analysis of the average number of positive cells per section in area 24, the cruciate gyrus and area 23 in the parasplenial gyrus of both formalin-injected group and control group was performed with the Mann-Whitney U Test. The average numbers of positive cells in areas 24 and 23 (means, 97.4 and 68.1, respectively) for the experimental group (N = 4) were more than double those means of control group (N = 2)(44.9 and 9.0, respectively), but this difference was not significant (p = 0.16, 0.06, respectively). The corresponding number for the cruciate gyrus in the experimental group (mean = 104.1) was not significantly higher than that (mean = 53.3) of the control group (p = 0.64).

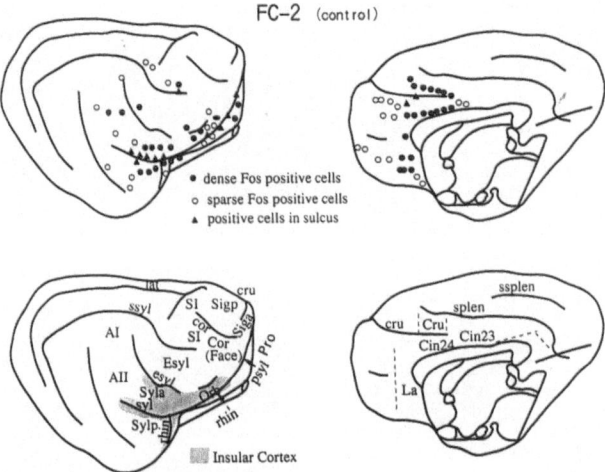

Fig. 2. Cortical distribution of Fos positive neurons in control cat FC-2 which received no injection but did undergo urethane anesthesia (upper figures). The distribution of Fos positive neurons was almost the same as that in cat FC-3 except for the absence of Fos neurons in the coronal gyrus. In the lower figures, the designations of gyri and sulci used in this study with Brodmann's area and functional areas are shown. *SI* primary somatosensory cortex; *AI, AII* primary and secondary auditory cortex, respectively

In the thalamus, no Fos positive cells were detected in the ventrobasal nucleus. Although Fos positive cells were commonly found in the paraventricular and midline nuclei, no difference was found between the two groups. In both groups Fos positive cells were also found in the hypothalamus including the paraventricular and supraoptic nuclei, PAG, locus caeruleus, parabrachial nucleus and other brainstem nuclei. The only difference between the two groups was the expression of c-fos in the trigeminal caudal nucleus in the experimental group (Fig. 3D).

Discussion

Anterior cingulotomy as advocated by Foltz and White [4] was not intended to destroy association fibers in cortex representing pain sensation, but to restore the normal functioning of the limbic system being disturbed by the presence of chronic pain. However, the findings of PET studies have suggested that the anterior cingulate cortex also participates in pain perception [11]. If this is the case, the interpretation of the effect of cingulotomy in patients with chronic pain must be revised.

The results of our retrograde labelling study indicate that sensory signals mediated by spinothalamic pathways are relayed to area 24 via thalamic midline and intralaminar nuclei, MD and possibly the submedial nucleus, each of which receives a contingent of the

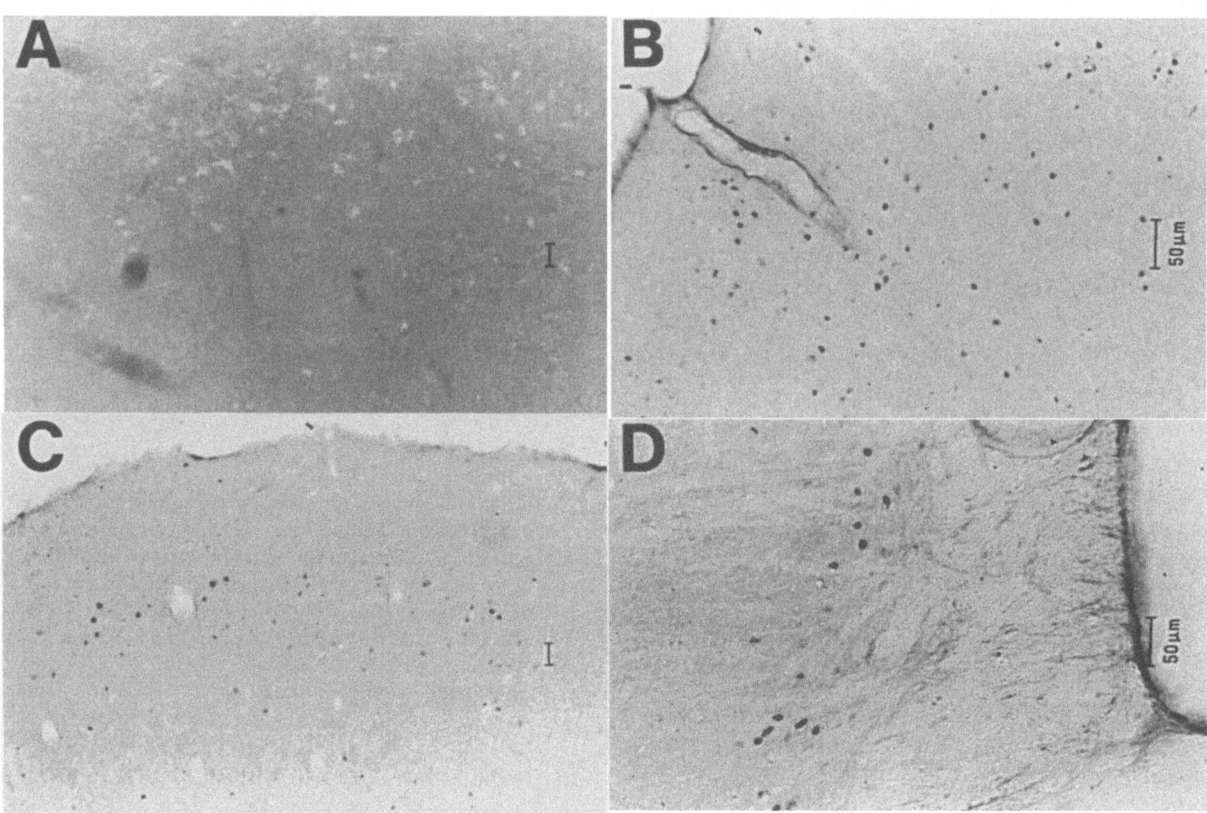

Fig. 3. Photomicrographs showing (A) retrogradely labelled cells in the medical central nucleus in cat 419; (B) Fos positive neurons in area 24 in a formalin-injected cat; (C) Fos positive neurons in the coronal gyrus in a formalin-injected cat; and (D) Fos positive cells in the the marginal layer of the trigeminal caudalis uncleus following formalin injection. Scale in each photograph: 50 μm

spinothalamic fibers [7]. The results of this WGA-HRP study thus suggest that, in fact, a portion of pain information is transmitted to area 24.

Although our findings suggested that the number of Fos positive neurons in the anterior cingulate cortex was increased 2 hours after formalin injection into the trigeminal territory, control animals exhibited a similar distribution of Fos immunoreactivity. Moreover, no significant difference in the number of Fos positive cells was observed between cats with and without formalin injection. If induction of anesthesia is associated with stress due to restraint, the similarity of Fos expression in experimental and control animals could be explained as a result of stress. In fact, various types of stress have been reported to increase c-fos expression in the anterior cingulate cortex [13], the paraventricular hypothalamic nucleus and in brainstem nuclei participating in brain circuits mediating strees reactions [1]. The main sites of c-fos expression in both groups coincided with these stress-related parts of the brain, except for the brainstem and terminals in the somatosensory cortex. In conclusion, it seems likely

that pain information reaches area 24 via the midline and intralaminar nuclei of the thalamus, following stress or emotional reactions. Surgical intervention in the anterior cingulate cortex such as cingulotomy therefore still seems to produce its effects by the interruption of circuits involved in stress and emotion.

References

1. Ceccatelli S, Villar MJ, Goldstein M, Hökfelt T (1989) Expression of c-Fos immunoreactivity in transmitter-characterized neurons after stress. Proc Natl Acad Sci USA 86: 9569–9573
2. Dragunow M, Faull R (1989) The use of c-fos as a metabolic marker in neuronal pathway tracing. J Neurosci Meth 29: 261–265
3. Fitzgerald M (1990) c-Fos and the changing face of pain. TINS 13: 439–440
4. Foltz EL, White LE (1962) Pain "relief" by frontal cingulumotomy. J Neurosurg 19: 89–100
5. Hassler R, Muhs-Clement K (1964) Architektonischer Aufbau des sensomotorischen und parietalen Cortex der Katze. J Hirnforsch 6: 377–422
6. Krettek JE, Price JL (1977) Projections from the amygdaloid complex to the cerebral cortex and thalamus in the rat and cat. J Comp Neurol 172: 687–722

7. Mantyh PW (1983) The spinothalamic tract in the primate: a re-examination using wheat germ agglutinin conjugated to horseradish peroxidase. Neuroscience 9: 847–862

8. Mesulam MM (1978) Tetramethyl benzidine for horseradish peroxidase neurohistochemistry: a non-carcinogenic blue reaction prouct with superior sensitivity for visualizing neural afferents and efferents. J Histochem Cytochem 26: 107–117

9. Musil SY, Olson CR (1988) Organization of cortical and subcortical projections to anterior cingulate cortex in the cat. J Comp Neurol 272: 203–218

10. Papez JW (1937) A propozed mechanism of emotion. Arch Neurol Psychiat 38: 725–743

11. Roland P (1992) Cortical representation of pain. TINS 15: 3–5

12. Rose JE, Woolsey CN (1948) Structure and relations of limbic cortex and anterior thalamic nuclei in rabbit and cat. J Comp Neurol 89: 279–347

13. Stone EA, Zhang Y, John S, Filer D, Bin G (1993) Effect of locus coeruleus lesion on c-fos expression in the cerebral cortex caused by yohimbine injection or stress. Brain Res 603: 181–185

14. Yasui Y, Ito K, Kamiya H, Ino T, Mizuno N (1988) Cingulate gyrus of the cat receives projection fibers from the thalamic region ventral to the ventral border of the ventrobasal complex. J Comp Neurol 274: 91–100

Correspondence: R. Kuroda, M. D., Department of Neurosurgery, Kinki University, Ohno-Higashi 377–2, Osaka-Sayama, Osaka 589, Japan.

Acta Neurochir (1995) [Suppl] 64: 74–78

Localisation of Epileptic Foci with Multichannel Magnetoencephalography, MEG

E. Knutsson and **L. Gransberg**

Department of Clinical Neurophysiology, Karolinska Hospital, Stockhom, Sweden

Summary

With the development of multichannel magnetoencephalographs biomagnetic signals can be recorded over large areas at the same time. It allows determination of the magnetic field outside the head generated by spontaneous epileptic discharges. From the maxima of outward and inward magnetic fluxes the locations of the sources of epileptic discharges can be calculated. The biomagnetic signals originating from an epileptic discharge is, however, mixed with biomagnetic signals generated by the background activity of the brain. A localisation based on a single discharge will therefore be influenced by the background activity. To diminish this influence, the biomagnetic signals during repeated identical epileptic discharges have been averaged. It gives a source localisation common to all discharges instead of a widely spread cluster of dipole sources. The error of epileptic dipole source localisations varies with the site of the dipole in the head as judged from studies with implanted dipole sources but also with the direction of the dipole as seen in studies with artificial dipoles in real head volumes. The error is relatively small in areas where the head has a spherical surface if the dipole direction is tangential. At other sites or dipole directions, the error can be very large. Since the site and direction of an epileptic dipole source is unknown it is not possible to estimate the error of localisations except by using models of individual head volumes.

Keywords: Epileptic foci; magnetoencephalography; EEG.

Introduction

Magnetic signals outside the head generated by brain activity was first recorded in the late sixties. Using a magnetometer placed close to the surface of the head, Cohen [2] recorded biomagnetic signals originating from alpha activity. Magnetoencephalographic records were at that time taken with relatively primitive instruments based on the Josephson phenomenon found a few years earlier [4]. The phenomenon implies that electrons can pass from one supraconducting region to another separated by a resistive barrier, a weak link. This mechanism is used in the sensing element of the Supraconducting QUantum Interference Device (SQUID) used in magnetometers to measure magnetic flux. The magnetometer consists of a detection coil connected to a SQUID that converts magnetic flux to an electric signal. It has to be kept at a sufficiently low temperature for supraconduction, *i.e.*, at $-269\,°C$, attained by having the device immersed in liquid helium.

The localisation of brain activity with magnetoencephalography is based on the fact that an electric current in the brain gives rise to a magnetic field outside the brain. The current generated by a single activated neurone is too weak to give a magnetic field that can be recorded but if a large group of neurones is activated at the same time, the resulting magnetic field may be so large that it can be recorded at the outside of the brain. It seems likely that about 10,000 neurones have to be activated in synchrony to give biomagnetic signals sufficiently large to allow their identification with modern instruments.

Magnetoencephalographs

The first magnetometers had a single detection coil. It meant that only a small fraction of the magnetic field outside the head could be recorded at the same time. Magnetic field determinations are required to localise sources in the brain of biomagnetic signals. Thus, measurements with the detection coil in different positions had to be done while the same brain activity had to appear repeatedly [1,7]. In consequence, determinations of magnetic fields took very long time and the identity of repeated spontaneous activity like, *e.g.*, epileptic spikes, was hard to guarantee. In contrast, cortical evoked responses were easy to repeat in a well defined way. The magnetic fields they generated could therefore be determined more easily.

During the last decades, there has been a continuous methodological development of magnetoencepahlography with improved signal/noise ratios and increased number of detection coils for concomitant records of biomagnetic signals over successively larger areas of the head. Today there are instruments that can record over one side or two sides of the head and others that can record even over the entire head.

The magnetic fields generated by brain activity are very weak. Those generated by cortical evoked responses are often not larger than about 0.1 pT. Epileptic discharges usually give fields with maximum strength of 1–5 pT. Considering the facts that the background magnetic field at the surface of the earth is about 50,000 pT and that the field noise in common environments usually is larger, gives an idea of the problem involved in differentiation of the biomagnetic signals generated in the brain. In part it has been solved by using two or more coils in series to give a gradiometer in stead of a magnetometer. One coil is kept close to the head surface while the other or others are kept 5–10 cm from the head. Magnetic fields generated by distant sources will be of approximately the same strength at all coils. Thus, by having the coils coupled in series but with opposite directions, the magnetic flux from distant sources will outbalance each other. The biomagnetic signals generated in the brain will, however, be much stronger close to the head than more distant from it, and the difference will indicate a field generated in the brain. To further diminish the effect of environmental field noise, recordings are made in magnetically shielded rooms made with, e.g., three from each other isolated walls of aluminium, copper and mu-metal, respectively.

Localisation of dipole sources of epileptic signals and cortical evoked responses are less complex when based on MEG than when based on EEG recordings. In determinations of the sources of EEG signals, calculations have to include estimations of current distribution in at least three layers corresponding to the brain, the scull and the scalp. In contrast, determinations of the source of biomagnetic signals can be based upon calculations of the current spread within a single volume limited by the inside of the scull. Therefore, MEG has been used extensively for localisation of brain activities like cortical evoked potentials, slow waves and epileptiform brain activity.

Records of MEG and EEG in Focal Epilepsy

It is quite common that epileptogen biomagnetic signals are recorded even when EEG recordings do not show any epileptiform activity. Figure 1 gives an example that shows epileptiform biomagnetic signals in MEG though concomitant EEG did not reveal any epileptiform activity. It depends on the fact that the voltage difference between scalp electrodes recorded with EEG could not differentiate low voltage signals caused by epileptic discharges from the background EEG activity. If, however, series of EEG signals are averaged using the peaks of a series of equal MEG recorded "spikes" to synchronise the averaging, the background EEG activity successively becomes outbalanced while an epileptiform spike of low amplitude appears in the averaged EEG [5].

The possibility to distinguish an epileptiform signal against background activity in MEG and EEG depends on the strength of the epileptic discharge. This strength varies with the number of neurones synchronously activated and their distance from the detection coils or scalp electrodes. Thus, the fact that MEG recordings are more sensitive implies that they some-

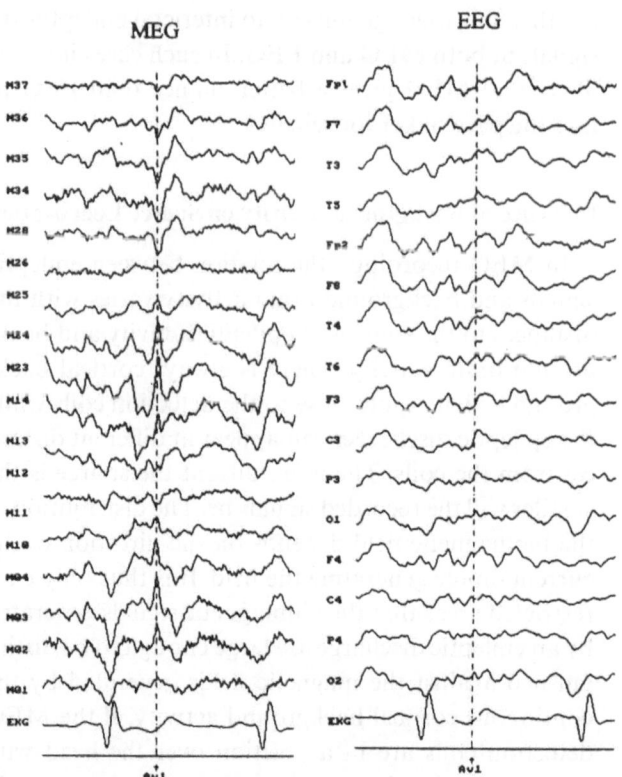

Fig. 1. Concomitant MEG and EEG recordings in patient with intractable focal epilepsy. Bidirectional, epileptiform signals in the MEG recordings with peaks indicated by vertical dotted line (Av1). No epileptiform signals in concomitant EEG. Lowest record is ECG

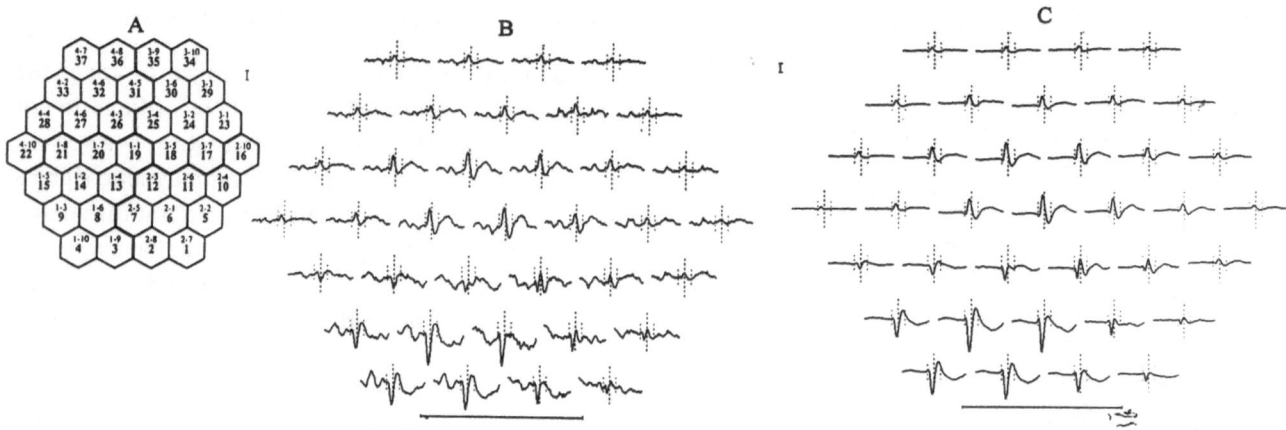

Fig. 2. (A) Organisation of the 37 detection coils of multichannel magnetoencephalograph. (B) Biomagnetic signals in the 37 channels during period with a single epileptic discharge. (C) Averaged MEG signals from 40 epileptic discharges of the type seen in (B). Vertical bars, 1 pT; horizontal bars, 1 s

times can reveal an epileptic discharge at an earlier stage of its development, *i.e.*, before so many neurones have become involved that it can be seen in EEG skalp recordings. Thus, in MEG but not concomitant EEG recordings we have seen interictal epileptiform signals with origin at one site that within a few milliseconds have grown larger and at the same time moved to another site, there giving rise to interictal epileptiform signals in both MEG and EEG. In such cases it seems likely that MEG gives a better chance than EEG to find the pacemaker for seizures.

Influence of Background Activity on Source Localisation

In MEG recordings, the relation between epileptic signals and background brain activity varies with the distances to the sources of epileptic activity and background brain activity. There is always cortical background activity quite close to the detection coils while the epileptic discharges can appear at different distances from the coils. The more distant the source is the smaller will the recorded signals be. The distribution of the biomagnetic field depends on the direction of the current dipole generating the field. It is thus only over restricted areas that the biomagnetic signals generated by an epileptic discharge are large enough to be distinguished against the magnetic fields generated by the continuous cortical background activity. If the MEG detection coils are in a position over the head with maximum epileptiform biomagnetic signals, the ratio between the amplitude of epileptiform signals and background activity is commonly larger in MEG than in scalp EEG recordings. When they are in other

positions, the opposite is true, and epileptiform activity may be seen in scalp EEG but not in MEG.

Figure 2A shows the form and organisation of detection coils in the magnetoencephalograph (KRENIKON, Siemens) that we have used in studies of focal epilepsy. It has 37 gradiometers in a nearly circular plane (19 × 17 cm) with a distance of 27 mm between the centres of the coils. The biomagnetic signals recorded with the different gradiometers during a single, focal epileptic discharge in the left temporal area of a child is seen in B. In C, each record is an average of 40 biomagnetic signals synchronised to the peaks of spikes (dotted vertical lines) in the MEG recordings. As can be seen in the figure, the biomagnetic signals have opposite direction in two areas of the magnetic field indicating inward and outward magnetic flux. The source of the field is usually estimated to be under a line between the maxima of outward and inward magnetic flux on a distance from the plane of the detection coils equal to the distance between these maxima divided by the square root of 2.

The biomagnetic signals generated by an epileptic discharge are mixed with biomagnetic signals generated by the continuous background activity. When the epileptic discharge is repeated, its magnetic field will be mixed with the field from a different background activity. Therefore, the magnetic fields recorded during repetition of epileptic discharges can never be identical even if the epileptic discharges are identical. In localisations of dipole sources based on single epileptic discharges, influences from the cortical background activity can not be excluded. Since the main part of the background activity is generated close to the detection coils while an epileptic discharge may be generated at a more distant site its influence on the localisation can be quite large. A reduction of the influence of the background activity on localisation is obtained by averaging the biomagnetic signals generated by identical epileptic discharges. Identification of identical epileptic discharges is however not an easy task considering the fact that the mixture of epileptic signals with background activity never results in precisely the same set of biomagnetic signals.

With careful comparison of the biomagnetic signals generated by different epileptic discharges in displays similar to the one of Fig. 2B, virtually identical discharges seem possible to identify. To assess identity we used similar field distributions as criterion for identity considering the sites of maxima of outward and inward magnetic flux, signal amplitude and amplitude relations between series of signals while accepting differences only when they could be explained by differences in background activity. Figure 3 shows localisations of dipole sources for a series of 17 virtually identical epileptic discharges. In A is shown localisations based on the biomagnetic fields generated by each one of the 17 discharges. In B, localisation was based on the averaged field from the same 17 discharges. The localizations based on single discharges show a wide dispersion with different dipole positions even within a single spike discharge. In contrast, the localisation based

upon the averaged field indicates a circumscribed source that did not change appreciably during the spike discharge.

Error of Localisations with Spherical Head Models

Localisations of dipole sources in the brain have generally been based on spherical models of the head. Though the human head is far from spherical, they seem to have an acceptable precision when the localisations are based on biomagnetic signals over spherical parts of the head if the dipole source has a tangential direction. When based on signals over non-spherical parts of the head or when the dipole direction is not tangential, the localisations are far from correct. Studies of dipole localisation in the human head with implanted electrodes have indicated error 1–4 mm [8] up to 20 mm [9]. Our own tests of dipole localizations based on the spherical model using artificial dipoles of different directions in volumes corresponding to individual, real head volumes indicate large variation in error with position and direction of the dipole as well as with the form of the head. It seems likely that these errors can be reduced markedly by using models of the real head volume determined for each individual. Experiences with such models are, however, limited, and their precision has to be tested before their feasibility can be assessed.

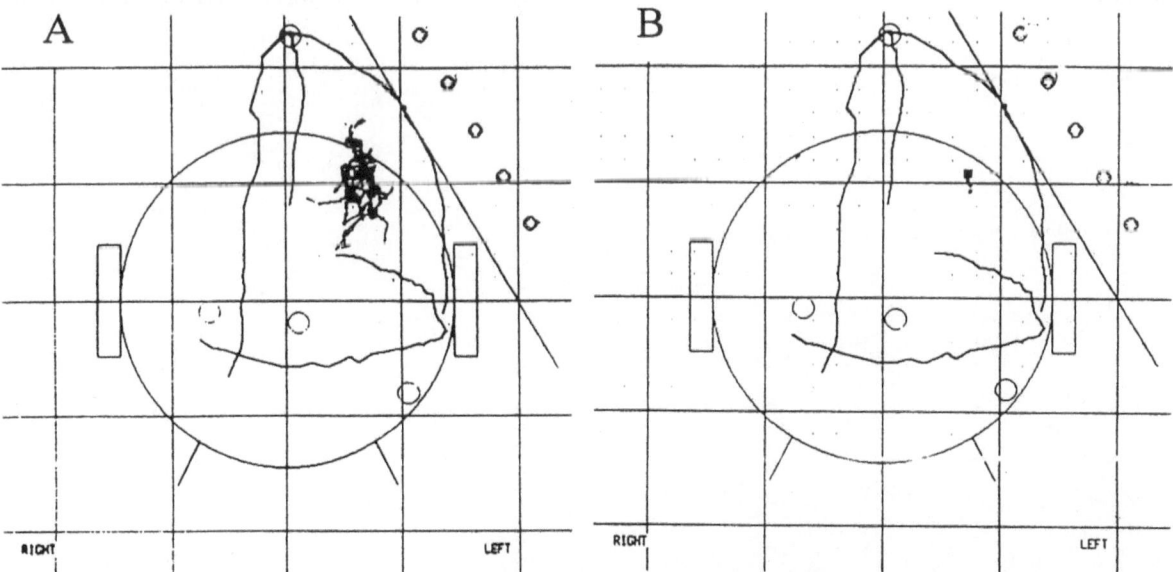

Fig. 3. (A) Projection from a frontal view of calculated dipole sources of 17 single, virtually identical epileptic discharges in patient with an epileptic focus in the left hemisphere. (B) Projection of the dipole source calculated from the mean of the same 17 epileptic discharges. Curved lines indicate surface of the head in an anteroposterior direction, and laterally in horizontal and ventrical planes. Planes of the magnetoencephalograph surface and of the detection coils indicated by lines and rows of circels to the right. Sides of squares, 50 mm

MEG in Preoperative Analysis of Intractable Epilepsies

So far, MEG localisation of the sources of interictal epileptic activity has been used in preoperative analysis of intractable focal epilepsy to indicate areas from which frequent interictal epileptic discharges are generated. It seems that clusters of sources of epileptic activity calculated from magnetic fields of interictal spike discharges usually are within the area from which seizures start as determined from studies of seizures with subdural electrodes [3,6] and occasional seizures during MEG recordings [10].

The finding that epilepsy in patients with vascular malformations usually disappears after radiosurgery that only affects a small, circumscribed volume [11] led to the question whether circumscribed gamma-knife lesions could be used to control focal, intractable epilepsy. So far, gamma-lesions have been used in a few patients with epileptic activity generated from sources calculated to be closely associated to structural abnormalities with effects on seizures comparable to those after open surgery. With a new approach, it seems relevant to select patients with frequent, interictal epileptic activity from only one area and verify its source localisations in models of the real head volume. Then the verified source localisation can be used to optimise placement of depth electrodes to find an area from which seizures start. This approach made possible with multichannel magnetoencephalography has been found very promising.

References

1. Barth DS, Sutherling WW, Engel J Jr, Beatty J (1982) Neuromagnetic localization of epileptiform spike activity in the human brain. Science 218: 891–894
2. Cohen D (1968) Magnetoencephalography: detection of magnetic fields produced by alpha rhythm currents. Science 161: 778–786
3. Hellstrand E. Abraham-Fuchs K, Jernberg B, Kihlström L, Knutsson E, Lindqvist C, Sneider S, Wirth A (1993) MEG localization of interictal epileptic focal activity and concomitant stereotactic radiosurgery. A new non-invasive approach for patients with focal epilepsy. Physiol Meas 14: 1–6
4. Josephson BD (1962) Possible new effect in superconducting tunnelling. Phys Lett 1: 251–253
5. Knutsson E, Gransberg L, Nergårdh A, Hellstrand E, Åmark P (1994) Epileptiform biomagnetic signals in MEG in children with normal EEG. In: Proceedings VIIth European Congress of Clinical Neurophysiology, Budapest, p 10
6. Knutsson E, Hellstrand E, Schneider S, Striebel W (1993) Multichannel magnetoencephalography for localization of epileptogenic activity in intractable epilepsies. IEEE Transact Magnet 29: 3321–3324
7. Modena I, Ricci GB, Barbanera S, Romani GL, Carelli P (1982) Biomagnetic measurements of spontaneous brain activity in epileptic patients. EEG Clin Neurophysiol 54: 622–628
8. Sato S (1990) Epilepsy research. NIH experience. In: Sato S (ed) Magnetoencephalography. Raven, New York, pp 223–230
9. Smith DB, Sidman RD, Flanigin H, Henke J, Labiner D (1985) A reliable method for localizing deep intracranial sources of the EEG. Neurology 35: 1702–1707
10. Stefan H, Schneider S, Feistel H, Pawlik G, Schuler P, Abraham-Fuchs K, Schlegel T, Neubauer U, Huk W (1992) Ictal and interictal activity in partial epilepsy recorded with multichannel magnetoencephalography. Epilepsia 33: 874–887
11. Steiner L, Lindqvist C, Adler JR, Torner JC, Alves W, Steiner M (1992) Clinical outcome of radiosurgery for cerebral arteriovenous malformations. J Neurosurg 77: 1–8

Correspondence: E. Knutsson, M.D., Department of Clinical Neurophysiology, Karolinska Hospital, S-17176 Stockholm, Sweden.

Acta Neurochir (1995) [Suppl] 64: 79–82
© Springer-Verlag 1995

Partial Seizures with Onset in Central Area: Use of the Callosal Grid System for Localization

R. M. Lehman[1] and H.-I. Kim[2]

[1]UMDNJ – Robert Wood Johnson Medical School, New Brunswick, N. J., U.S.A., and [2]Chonbuk National University Medical School, Chonju, Korea

Summary

Focal seizures arising in the central area require precise anatomic and physiologic mapping of ictal onset. The central sulcus is identified by the callosal grid system whose mid-vertical plane identifies the central sulcus inferiority where the central artery passes into the central sulcus. 5 patients with intractable seizures of central origin where localized with this method. Extent of resection was confirmed on postoperative NRI. The subdural grid was accurately placed on the central sulcus, confirmed by electrophysiologic means. Grid planes compartmentalized ictal onset, and post-operative resection correlated precisely. All patients are seizure-free. Seizures arising in the central area, precisely located, can be treated with good to excellent results. Localization of onset is facilitated by use of the callosal grid system, and allows superimposition of pre-, intra- and post-operative anatomic and physiologic data.

Keywords: Partial seizures; central area; callosal grid localization.

Introduction

The results of extratemporal surgery for the relief of intractable seizures of focal onset are inferior to temporal lobectomy [8]. Extent of resection and eloquence of cortex are the major factors determining this outcome. However, focal seizures arising in the central area can be treated surgically with good results [4]. This requires precise localization facilitated by use of subdural grid for both intracranial recording and functional mapping. Pathology, extent of lesion, and ictal activity resection are all determinants for assessing clinical outcome [10]. In order to correlate the extent of resection, as determined with postoperative MRI, with the intra-operative determination of the required resection, it is mondatory to coordinate the physiological data with precise anatomical localization. To this point, the central sulcus is the focus of our initial anatomical mapping in patients having partial seizures in the central area. We describe the use of the callosal grid system, to facilitate precise anatomical mapping based on the relationship of the midcallosal plane to the central sulcus [5].

Patients and Methods

Five patients with focal epilepsy of varying pathology arising in the central area, were studied with intracranial monitoring, using subdural grid and strip electrodes. In all cases a 64 contact grid was used in addition to supplemental strip electrodes. Monopolar and bipolar recordings were obtained, and functional mapping of the cortex for confirmation of localization of electrode contacts was performed using bipolar stimulation up to 10 ma. at 50 Hz.

The callosal grid system is based on the stereoscopic arteriovenous digital subtraction angiography (DSA) and magnetic resonance imaging (MRI). The corpus callosum can be seen indirectly on the arteriovenous DSA, and directly on the mid sagittal MRI (Fig.1). Imaging in either modality allows superimposition of identical landmarks. A proportional grid system is created by establishing a horizontal plane (HP) through the inferior border of the genu and the splenium of the corpus callosum. Three vertical planes perpendicular to HP are constructed: (i) an anterior callosal plane (AC), tangential to the anterior border of the genu; (ii) a posterior callosal plane (PC), tangential to the posterior order of the splenium; (iii) a midcallosal plane (MC), which corresponds to the midpoint between AC and PC (Fig. 2). Our studies have shown that the midcallosal plane intersects the central sulcus at a point where the central artery enters the central sulcus. The distal central sulcus then courses between planes MC and PC(3). Using this proportional grid system, the gyral and sulcal pattern on the preoperative MR images (Siemens 1.0 Tesla Magnetron SP 4000, version A2.5) were examined in relation to the central sulcus and the pathology so as to recognize these structures intraoperatively, and allow an assessment of the extent of resection in relation to the lesion and electrocorticography (ECoG) abnormalities on the postoperative MRI.

Results

Use of the callosal grid system allowed prompt recognition of the central area structures and associated pathology on the pre-operative MRI of patients

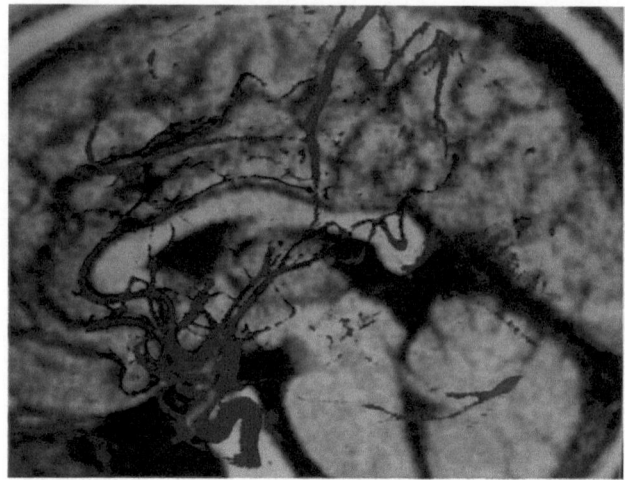

Fig. 1. Identification of corpus callosum superimposed DSA and MRI

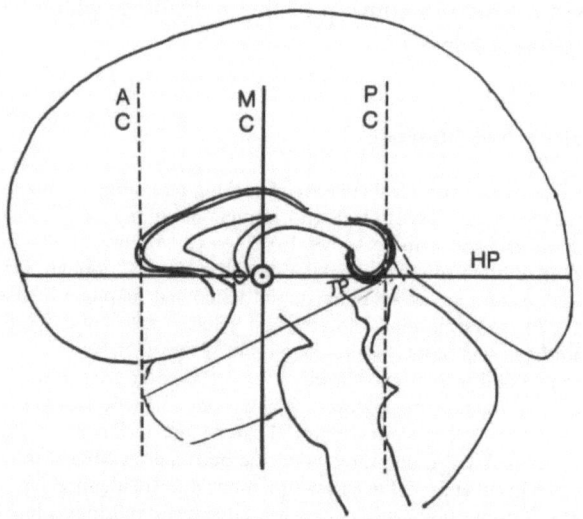

Fig. 2. Callosal grid system

with partial seizures in the central area. At the time of surgery the precentral and postcentral gyri were confirmed with cortical stimulation. The specific contact's location relative to the central sulcus was noted in placing the subdural grid. When consistent ictal data was recorded, a resection of the lesion and epileptogenic cortex was performed. Postoperative extent of the lesional and epileptogenic resection relative to the central sulcus and location of ictal contacts was recognized on MRI using the callosal grid system (Fig. 3). Sufficient removal of both lesion and epileptogenic tissue rendered four of five patients seizure free. The fifth patient will require a second stage removal in order to further reduce seizures (Table 1).

Discussion

Focal seizures arising in the central area are amenable to successful surgical treatment [13]. The preoperative anatomical mapping with modern imaging and intraoperative functional localization, coordinated with ECoG data, permits the resection of epileptogenic neocortex. A review of postoperative MR images with a clear recognition of the central sulcus, facilitated by the callosal grid system, allows the comparison with pre-operative MR images and the intra-operative ECoG data. This analysis allows the surgeon to reflect on the true extent of epileptogenic and lesional tissue. Mass effect in the central area will distort sulci and gyri. Their displacement within the grid vertical planes will predict their location at the time of surgery. Recently, functional MRI has been described, although length of time to perform the study and reproducibility of evoked images are not consistent [7]. Nonetheless, significant deficits may occur and

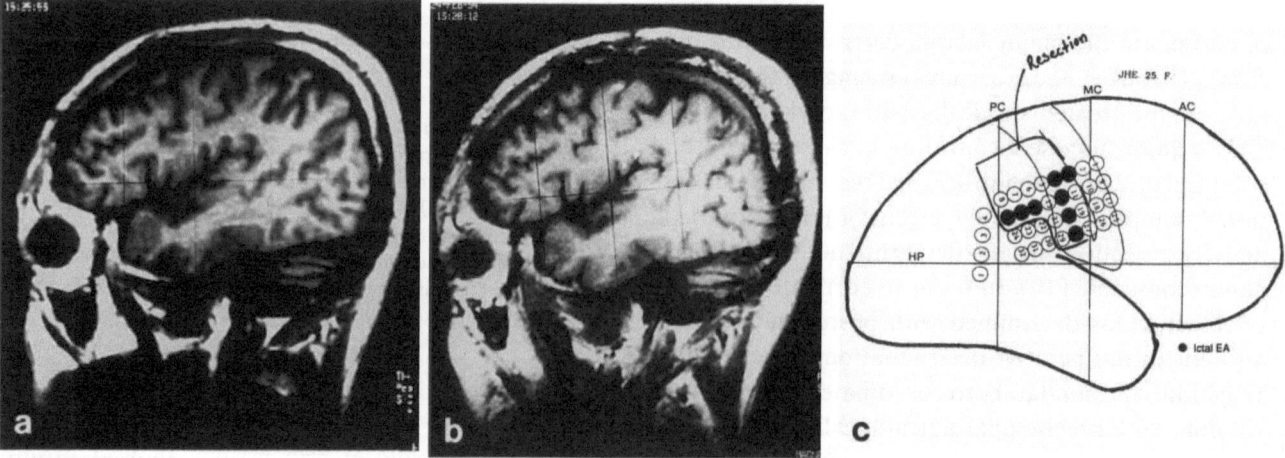

Fig. 3(a–c). Pre- and post-MRI images with location of central sulcus between MC and PC. Note location of subdural ictal contacts and resection on post-op MRI

Table 1

Pat.	Sex, age surgery	Age onset seizure	Clinical pattern	MRI abnor-mality	Area of ictal activity	Resection	Pathology	Result
LSM	F-17 yrs.	10 years	focal sensory and motor left leg and arm	bilateral centroparie-tal macrogyri	right post central rolandic, parasagittal	same area as ictal activity	cortical dysplasia	no seizure
JHE	M-25 yrs.	11 yeras	focal sensory then motor seizure left hand and arm GTC	post-central and supramar-ginal gyrus	limited in same area	same area as ictal activity	collagenoma	no seizure
SUZ	M-32 yrs.	5 years 1 year after en-cephalitis	left hemipariesis; aura of weakness left arm; followed by focal motor seizure GTC	abnormal-ity post central and pre motor	right post central and supramarginal gyrus	same area and additional anterior ictal activity	post encephaltic	no seizure
KEK	F-13 yrs.	8 years	paralysis left hand focal motor left face	insular abnormal-ity right	sensory more than motor face area	same area as ictal activity	non-specific changes	no seizure
CSY	M-2 yrs.	1 year 10 months	adversive eyes GTC	multiple cortical dysplasia	face area pre central pre motor supplimentary motor	face area pre motor 1st op	cortical dysplasia	no worthwhile results

GTC generalized tonic, clonic seizure.

seizures persist [9]. The location of the central artery and adjacent vascular structures to the central sulcus must be recognized pre and preoperatively, and preserved in their pial compartments in order to prevent remote injury in the central area [12]. Morrel and Whisler have proposed a technique of subpial resection to horizontally disconnect 5 mm vertical slices of cortex to address both of the problems of deficit and reduction of seizure activity [6].

Focal neuronal migration disorder and intractable partial epilepsy arising in the central area have been described [1, 2]. Extent of lesional resection correlated with reduction of seizures. More recently a unique aspect of ECoG data in focal cortical dysplastic lesion has been demonstrated by Palmini *et al.*: ictal-like electrographic activity recorded from dysplastic tissue. Lack of resection of this tissue correlates with a poor outcome [11]. Additionally, severity of histopathology and lack of lesion resection adversely affected seizure reduction. For successful resection it is crucial to localize not only the pathologic lesion but the electrographic abnormality, specifically the ictal-like activity

seen in focal cortical dysplasia. Patient 1 displayed this ictal-like activity, which was resected at surgery with resultant seizure-free postoperative condition. The callosal grid system applied to the pre and postoperative images with superimposed electrical data localizes this ictal and ictal-like activity and enhances the surgeon's recognition of location at the time of surgery and confirms its resection. In the event ictal activity is not resected and seizures continue, review of postoperative MRI images, using the callosal grid system, will reveal the extent of additional resection required and adjacent neurovascular structures. Our patients 2–4 with other pathology in the central area had seizure-free outcomes using the same evaluation and coordination of anatomical and functional data.

Conclusion

The callosal grid anatomical mapping methodology, coupled with the ability to localize and compartmentalize the abnormal electrographic activity within the subdivisions of the grid system, gives the surgeon comprehensive insight into the limits of lesional and electrographic resection required to achieve a seizure-free or marked reduction of seizure state for patients with intractable partial seizures arising in the central area.

References

1. Andermann F, Olivier A, Melanson D, Robitaille Y (1987) Epilepsy due to focal cortical dysplasia with macrogyria and the form fruste of tuberous sclerosis: a study of 15 patients. In: Wolf P, Dam M, Janz D, Driefuss FE (eds) Advances in epileptology, Vol 16. Raven, New York, pp 35–38
2. Kuzniecky R, Berkovic S, Andermann F, Melanson D, Olivier A, Leppik I (1988) Focal cortical myoclonus and rolandic cortical dysplasia: clarification by MRI. Ann Neurol 23: 317–325
3. Lehman R, et al (1992) Use of the callosal grid system for the preoperative identification of the central sulcus. Stereotact Funct Neurosurg 58: 179–188
4. Lenman RM, et al (1994) Seizure with onset in the sensorimotor face area: clinical patterns and results of surgical treatment in 20 patients. Epilepsia 35: 1117–1120.
5. Lehman R, Olivier A (1994) Use of the callosal grid system for cortical localization. In: Pasqualin A, DaPian R (eds) New trends in management of cerebro-vascular malformations. Springer, Wien New York, pp 333–338
6. Morrel F, Whisler W, Bleck T (1989) Multiple subpial transection: a new approach to the surgical treatment of focal epilepsy. J Neurosurg 70: 231–239
7. Mueller WM, Morris GL, Samanta-Roy R, Zerrin Yerkin F, Hammeke TA, Rao SM, Binder J, Wong EC, Bandettini P, Haughton VM, Hyde JS (1993) Cortical localization with magnetic resonance imaging compared to direct stimulation mapping. In: Meeting of the World Society for Stereotactic and Functional Neurosurgery. Abstract
8. Olivier A (1992) Extratemporal cortical resections. In: Luders H (ed) Epilepsy surgery. Raven, New York, pp 559–562
9. Olivier A (1991) Extratemporal resections in the surgical treatment of epilepsy. In: Spencer S, Spencer D (eds) Surgery for epilepsy. Blackwell, pp 150–167
10. Palmini A, et al (1991) Focal neuronal migration disorders and intractable partial epilepsy: a study of 30 patients. Ann Neurol 30: 750–757
11. Palmini A, Gambardella A, Andermann F, Dubeau F, Costa da Costa J, Olivier A, Tampieri D, Robitaille Y, Paglioli E, Neto EP, Coutinho L, Kim HI (1994) Operative strategies for patients with cortical dysplastic lesions and intractable epilepsy. Epilepsia 6: S57–S71
12. Rasmussen T (1975a) Cortical resection in the treatment of focal epilepsy. In: Purpura DP, Penry JK, Walter RD (eds) Advances in neurology, Vol 8. Raven, New York, pp 139–154
13. Rasmussen T (1975b) Surgery for epilepsy arising in regions other than the temporal or frontal lobes. In: Purpura DP, Penry JK, Walter RD (eds) Advances in neurology, Vol 8. Raven, New York, pp 207–226

Correspondence: Richard M. Lehman, M. D., Division of Neurosurgery, UMDNJ – Robert Wood Johnson Medical School, 1 Robert Wood Johnson Place, CN-19, New Brunswick, New Jersey 08903, U.S.A.

Acta Neurochir (1995) [Suppl] 64: 83–87
© Springer-Verlag 1995

Pain—Old and New Methods of Study and Treatment

W.H. Sweet

Massachusetts General Hospital, Boston, MA, U.S.A.

Summary

In the last 5 years several remarkable methods for localizing precisely a wide range not only of specific motor and sensory functions but as well of more complex mental phenomena in the domain of cognitive functions have been demonstrated to evoke sharply localizable responses. In pain, positron emission tomography (PET) scanning has been used to show that the anterior gyrus cinguli is an integral component of the pain system.

The PET technique suffers from a limitation of both spatial and temporal resolution, which permits only accurate center of mass coordinates of activated regions. Functional mapping of the brain by nuclear magnetic resonance has been achieved with techniques depicting specific brain areas in action during a mental process. These techniques open up an entirely new domain for study and treatment of many problems linked to cognition including many in whom pain is a central feature.

The many cerebral cortical areas involved in pain make it unlikely that any ablative procedure will achieve long sustained pain relief. The dual objective of relief of both pain and suffering is probably going to be attained only by activation of pain suppressor mechanisms. This may well require the added knowledge accessible only by functional magnetic resonance imaging.

Keywords: Pain; psychosurgery; functional MR imaging; cognitive process.

A. Leucotomy for Pain

In the early years of destructive operations on the frontal lobes for the treatment of otherwise intractable psychoses it soon became apparent that a variety of pains in the intractable category were also stopped for varying intervals by either relatively complete mid-frontal sections of white matter in the coronal plane or by a much more restricted incision bilaterally of the medial frontal white matter. However, lesions of major extent produced an unacceptable number of cognitive and behavioural sequelae especially among the large psychotic group in whom many hundreds of patients were treated in wholesale lots in institutions for the chronically psychotic. For this group the tactic of Freeman elicited the most vigorous criticism [7,22]. The saga of psychiatric surgery in the USA includes a happier chapter involving principally Hans Lukas Teuber and H. T. Ballantine. Teuber's distinguished contributions included the development of many new methods for demonstrating and measuring a wide variety of cognitive functions theretofore unappreciated [24]. In the ensuing 10 years a small but steady stream of carefully selected patients has continued to benefit from this operation of Ballantine.

Ballantine has also used his anterior cingulotomy for patients in pain. He has reported a group of 35 cancer patients in whom 25 lived less than 3 months [2]. His operation gave 57% of this short-lived group enough relief so that at most they required only non-narcotic analgesics plus other psychotropic drugs. Of the 10 living more than 3 months this degree of relief was achieved in only 2 of them. However, his results in a difficult group, 62 patients with failure to control pain continuing despite low back operations, are most encouraging. After a follow-up extending up to 7 years; 26% described 70–100% relief and 36% had 40–60% relief. In the group of 6 patients followed with post-operative chronic abdominal pain and the 5 with phantom pain 73% had 40–100% relief.

A single pair of inferior radiofrequency cylindrical lesions about 1 cm in diameter and 1 cm high placed in the subcaudate region (Fig. 1) with other pairs of lesions lateral and/or superior to the first lesion were used by Sweet and Poletti [23]. A varying number of lesions proved necessary to control the pain of 35 cancer patients. A single medial lesion was tried in 3 very ill patients. Table 1 describes the requirements of the remainder. Table 2 summarizes the results. Mentation and affect remained satisfactory in all but 2 patients in whom 3 pairs of lesions were made.

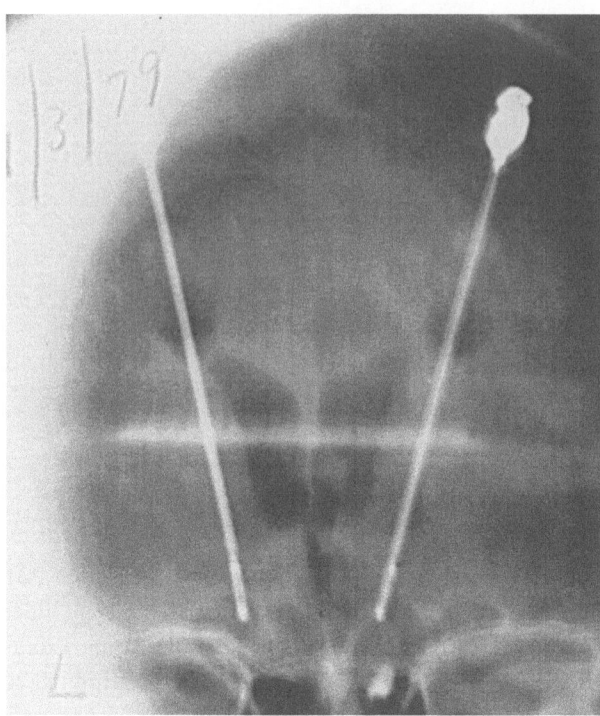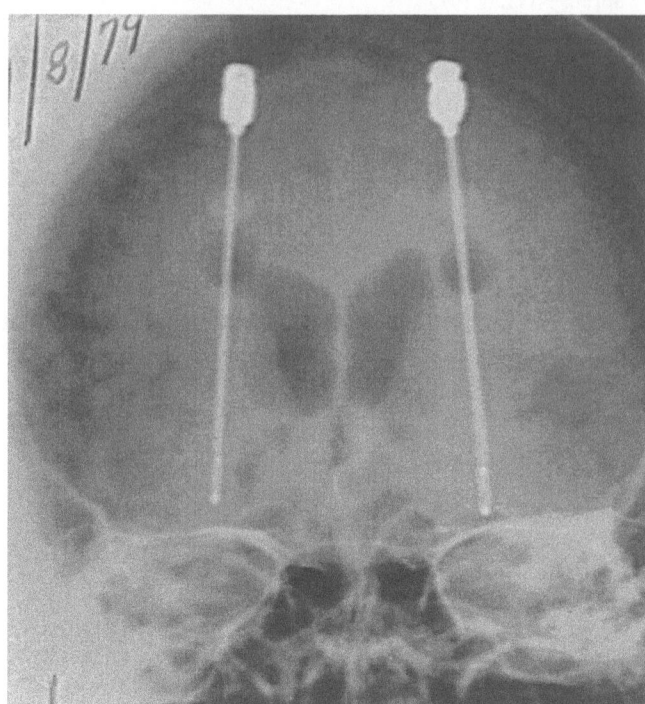

Fig. 1. Radiographs following placement of burr holes near the midline of the anterior hair-covered scalp via which air in the pre-CT scanning days was introduced to guide electrode placement for medial and lateral pairs of lesions in the preinnominate-subcaudate area. (Courtesy of Sweet *et al.* 1982) [23]

Table 1. *Number of Lesions in 35 Cancer Patients* [*23*]

Single medial lesion in 3 patients	
31 patients: status re pain assessable for months	
1 pair medial lesions	17
Medial and lateral pairs	8
6 lesions (3 pairs)	3
3 or 5 lesions	4

Table 2. *Results Re-Pain of Subcaudate Pre-Innominate Lesions.* With only one lesion 2 of 3 had relief till death in 1–2 weeks [23]

22 Patients	good to excellent relief after last of lesions
17 Patients	relief months till death or stupor
5 (of 22)	patients lost to late follow-up

B. New Methods for Studying and Treating Pain and Suffering

Functional Cerebral Mapping

Neurophysiologists and neurosurgeons have been able to determine the locus of detailed pain activator and pain suppressor pathways as far rostrally as the thalamus. In addition, electrical stimulation of cortical somato-sensory areas I and II and nearby areas in primary and secondary motor and supplementary sensory cortex may produce pain in patients with chronic pain which reproduces the clinical pain in the affected parts. However, we do not know the details of how this localized pain and the vaguer psychic suffering are generated or allayed. For example, we don't know how localized frontal leucotomies effect useful relief when they do so [8].

In the last 5 years several remarkable methods for localizing precisely a wide range not only of specific motor and sensory functions but as well of more complex mental phenomena in the domain of cognitive functions have been demonstrated to evoke sharply localizable responses. Some of those interested in pain fairly promptly used positron emission tomography (PET) scanning to show that the anterior gyrus cinguli is an integral component of the pain system.

This and other tactics for functional mapping depend at least partly upon the sharply localized changes in cerebral vascular hemodynamics which accompany local changes in metabolism. Happily the cerebral hemodynamic state in non-activated brain areas is relatively stable over time. The image of the perfusion area at rest may be subtracted from the image during specific activation to create a map localizing the function. For example, PET experiments demonstrate that sensory stimulation induces increases in cerebral blood flow and cerebral blood volume in the corresponding primary sensory cortex. In the PET imaging experiments that stimulation produced increases in regional cerebral blood flow without a commensurate increase in blood oxygen utilization pointing to an elevation in *venous* blood oxygenation [14].

Coghill *et al.* in a recent comprehensive summary (over 100 references) have depicted regional cerebral blood flow in PET scans obtained by the ^{15}o-water-bolus method in response to painful stimuli [6]. These produced activation of cortex in primary and secondary SI and SII somatosensory, anterior cingulate, anterior insular, supplementary motor and thalamic areas. They point out that painful stimuli were especially effective in activating the *anterior* insula, a zone linked with both somatosensory and limbic systems. The many cerebral cortical areas involved in pain make it unlikely that any ablative procedure will achieve long sustained pain relief. The dual objective of relief of both pain and suffering is probably going to be attained only by activation of pain suppressor mechanisms. This may well require the added knowledge accessible only by functional magnetic resonance imaging.

Functional MR Imaging

The PET technique suffers from a limitation of both spatial and temporal resolution, which permits only accurate center of mass coordinates of activated regions. Contrasting with this disadvantage, nuclear magnetic resonance is a high resolution totally innocuous in vivo method which coupled with recent advantages in scanning, speed results in an image every few seconds. Functional mapping of the brain by nuclear magnetic resonance has been achieved by 2 techniques. One involves injection of a bolus of a paramagnetic contrast agent such as gadolinium diethylenetriamine penta-acetic acid (GdDTPA). As the bolus passes through the capillary bed of the brain it produces changes in magnetic susceptibility of local blood volumes [21]. The variations in susceptibility are recorded with a fast magnetic resonance sequence using a transverse T2 weighted time relaxation. One may use either echo planar imaging or a fast low angle shot (FLASH) [4,9].

The second method makes use of a major change in magnetic behaviour of hemoglobin which depends on its degree of oxygenation. HbO_2 is diamagnetic, i.e. it is feebly *repelled* by a strong magnet whereas when dissociated from oxygen (deoxyhemoglobin, Hb) it becomes paramagnetic, i.e. is feebly *attracted* by the poles of a magnet [5]. Pauling and Coryell (1936) were the first to show that the Hb deoxyhemoglobin is paramagnetic and that the magnetic property of blood depends on its degree of oxygenation [17]. It was half a century before the observation was utilized in the current fashion. In this second method the hemoglobin in the body replaces the injected Gd-DTPA as the contrast medium, eliminating the need for an initial control run required by the use of an external contrast medium. Rapid changes in local tissue oxygenation are accompanied by equally rapid changes in magnetic susceptability [5]. Different imaging strategies are employed to emphasize one or more of several different physical phenomena that accompany the hemodynamic changes. For example, when blood water takes up a large fraction of the total water in a single image voxel one can measure an increase of the blood water signal due to a blood oxygenation-level-dependent (BOLD) change in T2 relaxation. Local magnetic susceptibility changes produced by changes in the concentration of deoxyhemoglobin in veins are emphasized in gradient-echo MR images. This has become the most widely studied method of MRI functional imaging. Thus Ogawa *et al.* with a fast T2 weighted imaging sequence and a 4 Tesla magnet were able to measure precisely the increase in signal from the stimulated visual cortex in man [16]. Visual stimulation produced with their equipment a 5–20% transient increase in the signal intensity in the human visual cortex while the time of acquisition was reduced from 40 to 8 msec. Moreover, the visual signal increases were 3–6 times greater and those in the motor cortex 2–4 times greater than those secured with the 1.5 Tesla magnet by Bandettini; [3] and Kwong [13]. Despite their advantage of working with a 4 Tesla magnet they were not able fully to explain the mechanism responsible for an increase in the signal during stimulation. One must certainly be prepared to search for other insignia of glial neuronal activity than cerebral blood flow and volume [5]. Kwong at the same task did obtain a better temporal resolution of 2–3 seconds with 1.5 Tesla echo planar imaging (EPI) [13].

With the same Hb-based method Bandettini *et al.* were able to follow the effects of finger motion in the motor cortex [3]. They saw similar signal changes within the human primary motor (MI) area during a hand squeezing task.

Functional Mapping of Cognitive Processes

Rao *et al.* [19] undertook the much more imaginative task of studying not only actual but imagined movements of the fingers as had Roland *et al.* (see their Fig. 9) [20]. In general sample self paced assigned simultaneous tapping of fingers 2 to 5 on 1 side induced activity only in the contralateral primary motor cortex. The more complex task consisted of tapping individual fingers in the same repeated sequence of middle, little, ring and index fingers. This induced an amazing spread of activity to both sides – contralateral > ipsilateral primary motor cortex, bilateral also of the supplementary motor area and premotor cortex (Brodmann's area 6) and the contralateral sensory motor area, plus less constant zones anterior and posterior to them. Finally the volunteer subjects were asked to imagine that they were carrying out the complex finger movement task while actually suppressing all motor activity. There were 1 to 4 small newly active areas in each of 4 MR slices. The largest in each of the 4 sections was always at the midline, probably anterior to the supplementary motor area. Pre- and post-operative functional MR mapping of the cerebral cortex especially of the motor, sensory and of the language areas has become standard practice in a number of centers [10,15,18].

For those of us interested in cognitive mechanisms as possible active components of persistent pain of our patients the publication of Kim, Ugurbil and Strick provides the most relevant of the functional MR maps to data [12]. The title "Activation of a Cerebellar Output Nucleus During Cognitive Processing" covers a comparison between the areas of the cerebellum activated by 2 similar, yet very different tests. A peg board with a line of 9 holes is supplied with 4 pegs in the 4 holes at one end of the board. Task A is to move each peg one hole at a time, into the holes at the opposite end of the board. The directions as well as the movements are the acme of simplicity. Task B called the "Insanity task" employs the same single row of 9 holes, but there are now 4 red pegs in the 4 holes at one end, 4 blue pegs in the 4 holes at the other end and a single empty hole between the 2 groups of 4. The task is to move the 4 pegs of each color from one end of the board to the other in accordance with 3 rules: 1) only one peg may be moved at a time; 2) a peg may be moved to an adjacent open space or may jump an adjacent peg; 3) a peg may be moved forward, never backward. During the visibly guided task A, 6 of the 7 volunteers developed a small region of activation in the cerebellar

dentate nucleus; it was bilateral in 4 of the 6. The major requirement of the insanity task was to think out the solution; the eye and limb movements were trivial. In fact, none of the subjects solved the problem during the period of scanning. All 7 of them developed a large bilateral activation of the dentate nuclei during their effort to solve the problem. The "dramatic increase in the size of the dentate activation" during the thought-provoking task led the investigators to conclude that the "regions of the dentate nuclei involved in congnitive processing are distinct from those involved in the control of eye and limb movement". These pioneering articles are the first I know of to come close to demonstrating an objective substrate for an area of the brain involved when thinking is occurring. The 2 articles depicting specific brain areas in action during a mental process open up an entirely new domain for study and treatment of many problems linked to cognition including many in which pain is a central feature.

Acknowledgement

The author wishes to express his gratitude to the Neuro-Research Foundation for its support during the preparation of this manuscript.

References

1. Ball J, Klett CJ, Gresock CJ (1959) The veterans administration study of prefrontal lobotomy. J Clin Exp Psychopathol Q Rev Psychiatr Neurol 20: 205–217
2. Ballantine HT Jr, Girinus IE (1988) Treatment of intractable psychiatric illness and chronic pain by stereotactic cingulotomy. In: Schmidek HH *et al* (eds) Operative neurosurgical techniques. Indications, methods, and results. Grune and Stratton, Orlando, pp 1069–1075
3. Bandettini PA, Wong EC, Hinks RD *et al* (1992) Time course EPI of human brain function during task activation. MRM 25: 390–397
4. Belliveau JW, Kennedy DN, McKinstry RC *et al* (1991) Functional mapping of the human visual cortex by magnetic resonance imaging. Science 254: 716–719
5. Blamire AM, Ogawa S, Igurbil K *et al* (1992) Dynamic mapping of the human visual cortex by high-speed magnetic resonance imaging. Proc Natl Acad Sci 89: 11069–11073
6. Coghill RC, Talbot JD, Evans AC *et al* (1994) Distributed processing of pain and vibration by the human brain. J Neurosci 14: 4095–4108
7. Freeman W, Watts JW (1950) Psychosurgery in the treatment of mental disorders and intractable pain. Transorbital lobotomy. Thomas, Springfield, pp 51–61
8. Gybels J, Sweet WH (1989) The neurosurgical treatment of persistent pain. Karger, Basel, pp 254–256
9. Hasse A, Frahm H, Hanicke W *et al* (1986) FLASH imaging. Rapid NMR imaging using low flip angle pulses. J Magn Reson 67: 257–266
10. Jack CR, Thompson RM, Butts RK *et al* (1994) Sensory motor cortex: correlation of presurgical mapping with function

MR imaging and invasive cortical mapping. Radiology 190: 85–92

11. Kim S-G, Ashe J, Hendrich K *et al* (1993) Functional magnetic resonance imaging of motor cortex hemispheric asymmetry and handedness. Science 261: 615–616

12. Kim S-G, Ugurbil K, Strick PL (1994) Activation of a cerebellar output nucleus during cognitive processing. Science 265: 949–951

13. Kwong KK, Belliveau JW, Chesler DA (1992) Dynamic magnetic resonance imaging of human brain activity during primary sensory stimulation. Proc Natl Acad Sci 89: 5675–5679

14. Menon RS, Ogawa S, Kim S-G (1992) Functional brain mapping using magnetic resonance imaging. Invest Radiol 27: S47–S53

15. Morris GL, Mueller WM, Yetkin FZ *et al* (1994) Functional magnetic resonance imaging in partial epilepsy. Epilepsia 35: 1194–1198

16. Ogawa S, Tank DW, Menon R *et al* (1992) Intrinsic signal changes accompanying sensory stimulation: functional brain mapping with magnetic resonance imaging. Proc Natl Acad Sci 89: 5951–5955

17. Pauling L, Coryell CD (1936) The magnetic properties and structure of hemoglobin, Oxyhemoglobin and carbonoxyhemoglobin. Proc Natl Acad Sci 22: 210–216

18. Rao SM, Bandettini P, Wong PC *et al* (1992) Gradient-echo EPI demonstrates bilateral superior temporal gyrus activation during passive word presentation. Book of Abstracts, 11th Annual Meeting. Society for Magnetic Resonance in Medicine, Berlin, p 1827

19. Rao SM, Binder JR, Bandettini PA *et al* (1993) Functional magnetic resonance imaging of complex human movements. Neurology 43: 2311–2318

20. Roland PE, Larson B, Lassen NA *et al* (1980) Supplementary motor area and other cortical areas in organization of voluntary movements in man. J Neurophysiol 43: 118–136

21. Rosen BR, Belliveau JW, Vevea JM *et al* (1990) Perfusion imaging with NMR contrast agents. MRM 14: 249–265

22. Sweet WH (1973) Treatment of medically intractable mental disease by limited frontal leucotomy – justifiable? N Engl J Med 289: 1117–1118

23. Sweet WH, Poletti CE, Umansky F (1982) Neurosurgical techniques to control the pain of superior pulmonary sulcus and other tumors in this region. In: Bonica JJ *et al* (eds) Advances in pain research and therapy, Vol 4. Raven, New York, pp 211–233

24. Teuber H-L, Corkin SH Twitchell TE (1977) Study of cingulotomy in man: a summary. In: Sweet WH *et al* (eds) Neurosurgical treatment in psychiatry, pain, and epilepsy. University Park Press, Baltimore, pp 355–365

Correspondence: W. H. Sweet, M.D., D.Sc., Massachusetts General Hospital, 5 Longfellow Place, Suite 211, Boston, MA 02114, U.S.A.

Acta Neurochir (1995) [Suppl] 64: 88–91

CT-Guided Pain Procedures for Intractable Pain in Malignancy

Y. Kanpolat, S. Caglar, S. Akyar, and **C. Temiz**

Department of Neurosurgery, Ibni Sina (Avicenna) Hospital Ankara University, Faculty of Medicine, Ankara, Turkey

Summary

CT-guided stereotactic percutaneous destructive procedures, i.e. percutaneous cordotomy, trigeminal tractotomy, and extralemniscal myelotomy, have been routinely used for the treatment of localized intractable pain in malignancy since 1987. In 67 cases if local pain due to malignancy, CT guided percutaneous cordotomy was performed, and in 97% complete pain control was achieved. In 45 of these cases, a "selective cordotomy" was performed meaning that analgesia was produced only in the painful region of the body. CT guided trigeminal tractotomy was applied to a total of 19 cases in 5 of which pain had been caused by malignancy. The results were satisfactory. 12 cases, suffering from visceral pain due to malignancy, were treated by CT-guided extralemniscal myelotomy and in 10 cases pain relief was achieved.

Keywords: Cancer pain; CT-guidance; stereotaxy; percutaneous procedures.

Introduction

Mortality from malignancy ranges between 10–20% [24]. Pain is one of the most important problems in patients with malignant diseases. Pain is present at the early stages in between 5 and 10% of the cases, and it reaches at the terminal stages up to 90% [3, 20]. Despite the extensive usage of new pharmacological agents, approx. 20% of cancer patients, especially in the terminal stage, still suffer from disabling pain due to improper treatment [19, 20]. Also recently developed narcotic and non-narcotic analgesics sometimes fail to control completely cancer pain. Morphine pumps have been widely used for such patients in terminal stage. Morphine, however, can induce new problems previously not experienced by the patient.

Ablative neurosurgical methods are particularly effective for the treatment of localized cancer pain. These procedures were very popular in 1960's and 1970's [7, 13–15]. However, over the last 10 years, the role of the neurosurgeon in the treatment of pain has gradually lost importance [5, 21]. This paper is concerned with the further development of pain surgery performed with selective localization whereby a more efficient pain relief and improved quality of life can be achieved. The techniques can be referred to as CT-guided pain procedures which includes percutaneous cordotomy, trigeminal tractotomy and extralemniscal myelotomy applied as treatment of localized cancer pain [9–11].

Material and Methods

The surgical procedures are carried out in a CT unit. The patients are placed in supine position for percutaneous cordotomy, in prone position for trigeminal tractotomy and extralemniscal myelotomy. CT images (512×512 matrix) with 2 mm slice thickness were used. The quality of the image was enhanced by diminishing the diameter of the image formation area (Picker SX 1200 and Picker IQ, Picker International Cleveland, Ohio, USA). The operative techniques have previously been described in detail [9, 12].

A special electrode system is used in the procedures: The KCTE Kanpolat CT Electrode Kit (Radionics® Inc, Burlington, Massachusetts) is a temperature monitoring RF electrode system for extralemniscal myelotomy, percutaneous cordotomy and trigeminal tractotomy. KCTE kits contain 20–22 gauge thin wall needles specially designed to diminish the metallic artifact interference in the CT-image. The 1 mm straight electrode is positioned in the central cord area in extralemniscal myelotomy. The 2 mm straight and 2 mm curved electrodes are positioned in the lateral spinothalamic tract for percutaneous cordotomy and in the descending trigeminal tract for trigeminal tractotomy, respectively. This specially designed electrode has a 2 mm exposed tip with a slight curve positioning the tip 0.5 mm away from the axis.

CT Guided Percutaneous Cordotomy

The target of the procedure is the lateral spinothalamic tract located in the anterolateral region of the spinal cord at C_1-C_2 level (Fig. 1) [13,15]. This procedure is especially indicated for patients suffering from unilateral localized cancer pain, e.g., Pancoast's tumor, chest wall mesothelioma, unilateral pelvic carcinoma, sarcomas located in one extremity.

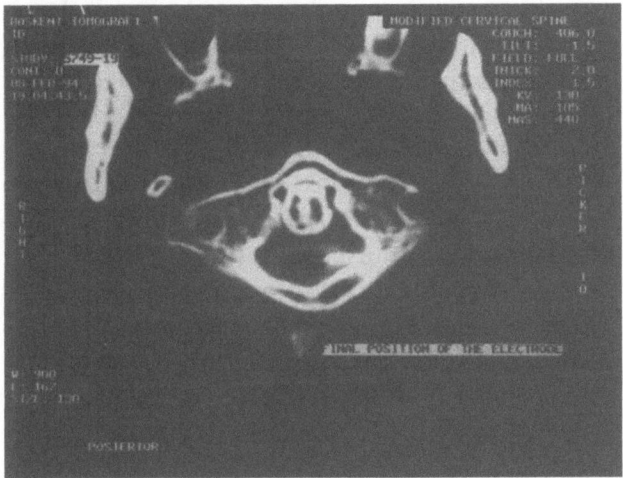

Fig. 1. Left anterolateral localization of the electrode system in the spinal cord at the C_1–C_2 level for CT-guided percutaneous cordotomy

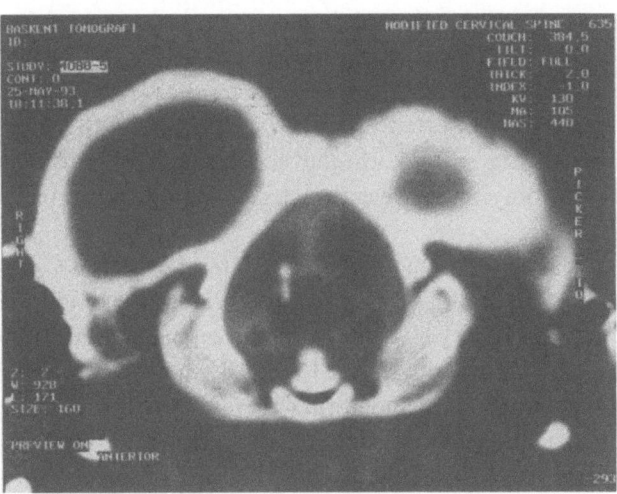

Fig. 2. Right posterolateral localization of the electrode system at the occiput-C_1 level for CT-guided trigeminal tractotomy

Pulmonary malignancies (33 cases) and mesotheliomas (6 cases) were the most common (58.2%) among the unilateral painful cases with malignancies (67 cases). Among the others, there were 7 gastrointestinal carcinomas, 5 osteogenic malignancies, 4 metastatic carcinomas, 3 genitourinary malignancies and 8 different types of several other malignancies.

CT-guided percutaneous cordotomy was applied to 66 cases unilaterally and to one bilaterally. In five cases, the procedure was applied twice because of failure of the treatment in two cases and early recurrence in the remaining three cases. Unilateral pain was relieved in 97% of the cases. In 45 cases (68%), only the painful region of the body was denervated. 4 cases who had been suffering from disabling pain returned to their work after the procedure. The complications were temporary ataxia in three cases and temporary hemiparesia in two. In four cases (three of them were Pancoast's tumors), pain due to malignancy was relieved but neuropathic pain related to the deafferentation recurred.

CT Guided Trigeminal Tractotomy

The target of this procedure is the descending trigeminal tract located in the posterolateral cervico-medullary portion of the spinal cord at the occiput-C_1 level (Fig. 2) [4, 6, 17, 18]. This procedure is particularly indicated for the treatment of unilateral cancer pain located in the areas of the V, VII, XI and X cranial nerves.

CT-guided trigeminal tractotomy was applied to 19 cases, the majority suffering from benign paroxysmal pain. 5 of them had malignant pathology (3 nasopharyngeal carcinoma, 1 parotid tumor, 1 metastatic breast carcinoma). In four cases with malignancy, intractable pain was controlled and in one, nucleus caudalis DREZ operation was performed because of insufficient pain relief. We did not observe any complications not even ataxia.

CT Guided Extralemniscal Myelotomy

The target of the procedure is the non-specific multisynaptic pathway located in the central cord region at the occiput-C_1 level (Fig. 3) [7, 16]. Patients suffering from visceral pain due to infiltrating malignancy of the abdomen and pelvis are ideal candidates for this procedure. This operation has been performed without any complications in 12 cases suffering from visceral pain due to malig-

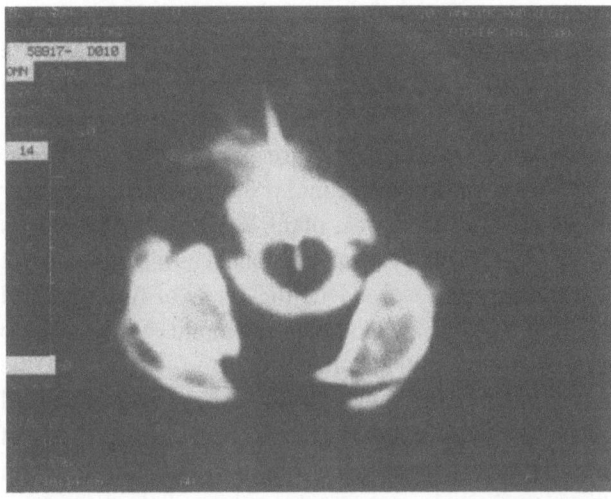

Fig. 3. Localization of the electrode system to the central cord area at the occiput-C_1 level for CT-guided extralemniscal myelotomy

nancy. Out of these cases, 3 suffered from rectal carcinoma, 3 gastric carcinoma, 2 colon carcinoma, 2 pancreatic carcinoma, 1 hepatoma and 1 renal cell carcinoma. Pain relief was achieved in 10 cases (4 total, 6 partial).

Discussion

In the last 20 years, destructive surgical pain procedures have lost a great deal of their popularity. The reason for this is not only the high efficacy of new methods like morphine pumps or neurostimulation but also neglect in the field of surgical pain procedures and electrode technology. CT-guided destructive pain surgery is not a new methodology. The important change compared with "blind" cordotomy is the use of the imaging CT technology. With the help of this system, the spinal cord diameter can be measured for each patient individually, the target-electrode relation can be assessed directly, and the needle electrode system can be inserted into the specific part of the pain conducting system for the achievement of selective cordotomy, and selective tractotomy, respectively [12, 16].

Selective destruction of the lateral spinothalamic tract was first described and performed by Hyndman and Van Epps [8]. Brihaye and Sweet had emphasized the possibility of selective lesions by open cordotomy, but had also concluded that there was a risk of pain recurrence due to incomplete dissection [2, 8, 22, 23]. In CT-guided cordotomy, destruction of a specific part of the lateral spinothalamic tract is possible. With the help of selective lesioning, it is not only possible to denervate the painful area but also to preserve other important tract functions. For this reason, percutaneous cordotomy was performed here with a 97% success rate and with very few complications. Selective precise lesion making carries a risk of recurrence of pain but this was observed in only three patients early after cordotomy, and the procedure was repeated with ease.

The descending trigeminal tract is an ideal target for denervation of pain areas of the V, VII, IX and X nerves [4, 6, 18]. Oropharyngeal, nasopharyngeal, and parotid tumours usually extend to the areas of these cranial nerves and cause intractable cancer pain. With the help of CT guidance, trigeminal tractotomy proved to be a very effective procedure for the denervation of painful areas supplied by these cranial nerves. In the case of neuropathic cancer pain, ablation of pain through tractotomy only can not be achieved. In that case, tractotomy has to be supplemented by a nucleus caudalis DREZ lesion [1].

Extralemniscal myelotomy was used for the treatment of abdominal and pelvic visceral intractable pain. This operation, however, was not extensively used by us because the underlying mechanism is poorly understood [7, 16].

Percutaneous cordotomy and trigeminal tractotomy are highly effective in controlling local intractable pain due to malignancy. After pain relief, most patients could discontinue the contact with the hospital and returned to a normal daily life at home. We consider this a major contribution to the well-being and psychology of the cancer patient.

Another important point seems to be controversial: Many physicians involved in pain practice consider destructive procedures only in the terminal stages of the malignant disease. We, however, advise destructive procedures, especially cordotomy and trigeminal tractotomy as early as possible. If the surgeons can denervate the painful area, the quality of life of the patient will improve and he will be in a position to carry on his normal active life.

In conclusion, CT-guided pain procedures for the treatment of intractable pain in malignancy are effective and selective. With the help of CT imaging and new electrode technology, these procedures can be used with more accuracy and selectivity. Neurosurgeons are encouraged to deploy this approach in dealing with pain in malignancy.

Acknowledgements

We would like to express our deepest appreciation to Mr. Ender-Arkun for his assistance in the preparation of this paper.

References

1. Bernard EJ, Nashold BS, Caputi E, *et al* (1987) Nucleus caudalis DREZ lesions for facial pain. Br J Neurosurg 1: 81–92
2. Brihaye J, Thiry S, Le Clerco R, Reti F, Gregoire A (1962) Le traitment chirurgical de la douleur. Acta Chir Belg [Suppl] 2: 255–475
3. Bonica JJ, Benedetti C (1986) Management of cancer pain. In: Moosa A, Kobson MC, Schmpff SC (eds) Textbook of oncology. William and Wilkins, Baltimore pp 443–447
4. Crue BL, Todd EM, Carregal EJ, Kilham O (1967) Percutaneous trigeminal tractotomy. Bull Los Ang Neurol Soc 32: 86–92
5. Gildenberg PL, Zanes C, Flitter MA, Lin PM, Lautsch EV (1969) Impedance measuring device for detection of penetration of the spinal cord in anterior percutaneous cervical cordotomy. Technical note. J Neurosurg 30: 87–92
6. Hitchcock ER (1970) Stereotactic trigeminal tractotomy. Ann Clin Res 2: 131–135
7. Hitchcock ER (1970) Stereotactic cervical myelotomy. J Neurol Neurosurg Psychiatry 33: 224–230
8. Hyndman OR, Van Epps C (1939) Possibility of differential section of the spinothalamic tract. A clinical and histologic study. Arch Surg 38: 1036–1053
9. Kanpolat Y, Atalag M, Deda H, Siva A (1988) CT-Guided extralemniscal myelotomy. Acta Neurochir (Wien) 91: 151–152
10. Kanpolat Y, Deda H, Akyar S, Bilgic S (1989) CT-Guided percutaneous cordotomy. Acta Neurochir (Wien) [Suppl] 46: 67–68
11. Kanpolat Y, Deda H, Akyar S, Caglar S, Bilgic S (1989) CT-Guided trigeminal tractotomy. Acta Neurochir (Wien) 100: 112–114
12. Kanpolat Y, Akyar S, Caglar S, Bilgic S (1993) CT-Guided percutaneous selective cordotomy. Acta Neurochir (Wien) 123: 92–97
13. Mullan S, Harper PV, Hekmatpanach J, Torres H, Dobbin G (1963) Percutaneous interruption of spinal pain tracts by means of a strontium 90 needle. J Neurosurg 20: 931–939
14. Mullan S, Hekmatpanach J, Dobbin G, Beckman F (1965) Percutaneous intramedullary cordotomy utilizing the unipolar anodal electrolytic lesion. J Neurosurg 22: 548–553
15. Rosomoff HL, *et al* (1965) Percutaneous radiofrequency cervical cordotomy. Technique. J Neurosurg 23: 639–644

16. Schvarcz JR (1976) Stereotactic extralemniscal myelotomy. J Neurol Neurosurg Psychiatry 39: 53–57

17. Schvarcz JR (1977) Post-herpetic craniofacial dysaesthesise; their management by stereotactic trigeminal nucleotomy. Acta Neurochir (Wien) 38: 65–72

18. Sjoqvist OP (1938) Studies on pain conduction in the trigeminal nerve: a conduction in the trigeminal nerve: a contribution to the surgical treatment of facial pain. Acta Psychiat Neurol [Suppl] 17: 1–139

19. Sundaresan N, DiGiacinto GV (1987) Antitumor and anti-nociceptive approaches to control cancer pain. Med Clin North Am 71: 329–348

20. Sunderesan N, Digiacinto GV, Hughes JEO (1989) Neurosurgery in the treatment of cancer pain. Cancer 63: 2365–2377

21. Taren JA, Ross D, Crosby JC (1969) Target physiologic corroboration in stereotactic cervical cordotomy. J Neurosurg 30: 569–584

22. Walker AE (1940) The spinothalamic tract in man. Arch Neurol Psychiatry (Chicago) 43: 284–298

23. White JC, Sweet WH (1969) Spinothalamic tractotomy. Pain and neurosurgeon. Thomas, Springfield, Illinois, pp 701–702

24. World Health Organization (1984) Cancer as a global problem. Weekly Epidemiological Records 59: 125–126

Correspondence: Y. Kanpolat, M. D., Department of Neurosurgery, Ibni Sina (Avicenna) Hospital, Ankara University, Inkilap Sok. 24/2, 06650 Kizilay, Ankara, Turkey.

Acta Neurochir (1995) [Suppl] 64: 92–96

Electrophysiological Monitoring During CT-Guided Percutaneous Cordotomy

M. Zileli[1], **E. Coşkun**[1], **I. Yegül**[2], and **M. Uyar**[2]

Departments of [1]Neurosurgery and [2]Algology, Ege University Faculty of Medicine, İzmir, Turkey

Summary

During percutaneous cordotomy, impedance monitoring and electric stimulation have been widely used to enable a precise localization of the lesion electrode. The purpose of this study was to examine the possibility that the usage of additional electrophysiological techniques could help in improving the precision of the placement of the lesion electrode.

Fourteen patients were monitored with 4 different techniques during CT-guided percutaneous cordotomy:
1) Median nerve somatosensory evoked potentials (mSEP): median nerve stimulation with recording from the scalp.
2) Spinal cord evoked potentials (SCEP): median nerve stimulation with recording via the cordotomy electrode.
3) Spinothalamic evoked potentials (SthEP): stimulation via the cordotomy electrode and recording from the scalp.

Ipsilateral and contralateral mSEPs and SCEPs did not change after the lesion. SthEPs showed a significant decrease in 10 of 12 patients provided the stimulus intensity was kept below that producing a motor response (approx 0.5–1 mA). There was no obvious relationship between changes of the evoked potentials and the clinical outcome of the cordotomy. Our results suggest that there may be a relationship between the extent of the lesion and the decrease of the spinothalamic evoked potentials.

Keywords: Percutaneous cordotomy; evoked potentials; spinal cord stimulation.

Introduction

Percutaneous cordotomy is still extensively used for the treatment of cancer related pain resistant to pharmacological treatment [1, 2, 7, 9–11]. For the correct location of the lesion electrode the following methods are often used: 1) CT-imaging using contrast agents [7,11]. 2) Impedance monitoring [10]. 3) Electric stimulation at lower frequencies (2 Hz) may produce ipsi-or contralateral motor responses. 4) Stimulation at higher frequencies (50 Hz) evokes a sensation of warmth or cold [6,9,10]. Since percutaneous cervical cordotomy enables the recording of electrical activity directly from the spinal cord we explored the possibility that the usage of different electrophysiological methods could increase the precision of the lesion electrode placement and further to examine if any changes of evoked activity could relate to the outcome of the procedure.

Material and Methods

Cordotomy Procedure

Percutaneous cordotomy was performed to 14 cancer patients who had inadequate pain control by conventional methods. There were two females and 12 male patients, 40–72 years of age. Clinical data as well as the outcome of cordotomy are summarized in Table 1.

Details of our surgical technique have been presented before [11]. The patients were awake but sedated. The subarachnoidal space was punctured at the level of C1-C2 with a 20 gauge cordotomy needle and 7–8 ml contrast medium (iohexol) was injected. Axial CT scans were performed and the spinal cord was punctured while monitoring of the impedance. Electric stimulation was performed at low (2 Hz) and high (50 Hz) frequency. The lesions were created with a Levin electrode attached to a lesion generator (Radionics RFG-3B) with heating for 10, 20 and 30 seconds at a temperature of 80 °C. The entire procedure was performed in the CT room of the Radiology Department.

Electrophysiological Monitoring

Once the lesion electrode was judged to have a satisfactory position in the anterolateral quadrant of the spinal cord it was connected to a 2 channel recording machine (Phasis, ESAOTA, Italy).

The following recordings were performed (Table 2): 1) *Median somatosensory evoked potentials* (mSEP): The median nerve at the wrist was stimulated electrically at 3.2 Hz with a pulse width of 0.2 ms using a constant current stimulator set at an intensity to evoked motor responses in the hand. The responses were recorded from the scalp with averaging technique.
2) Contralateral and ipsilateral *spinal cord evoked potentials* (SCEP) were evoked by median nerve stimulation and recording from the cordotomy electrode. The stimulation conditions were the same as used mSEP.
3) *Spinothalamic evoked potentials* (SthEP): Stimulation was applied via the cordotomy electrode and recordings were made from the scalp. The lesion electrode served as a cathode and the spinal needle

Table 1. *Clinical Features of the Patients*

n	Initials	Age/Sex	Diagnosis	Pain location	Pain dermatom	Lesion side	VAS pre	VAS post
1	HÇ	41 M	rectum CA	R coxae	T12–L1	L	10	2
2	HK	55 M	mesothelioma	R scapula	T2–4	L	10	3
3	EB	57 M	epidermoid CA	L shoulder	C4–5	R	10	2
4	MS	55 M	lung cancer	R neck, hip	L1–L2	L	10	5
5	SU	69 M	epidermoid CA	L body, chest	T1–T5	R	8	1
6	MS	72 M	epidermoid CA	R back	C4–T4	L	8	2
7	ŞŞ	40 F	breast cancer	R shoulder, arm	C5–C8	L	7	1
8	MK	60 F	lung cancer	L shoulder, arm	C5–T2	R	8	2
9	RC	47 M	adenocarcinoma	R arm	C6–T1	L	9	3
10	HK	57 M	lung cancer	R hip	T10–L2	L	8	3
11	SÖ	52 M	breast cancer	R shoulder, back	C5–T2	L	8	5
12	MP	60 M	lung (oat cell)	L shoulder, arm	C5–T1	R	6	1
13	OK	60 M	lung cancer	R body, chest	C6–T4	L	8	4
14	HB	66 M	lung (pancoast)	R shoulder, arm	C5–T1	L	8	2

as an anode. Stimulation intensity was chosen so that the threshold for a subjective sensation was assessed. Stimulation intensity was increased until muscle twitching appeared in the ipsilateral side (approx 2–3 mA). Scalp responses also to an intermediate stimulation intensity were recorded. Stimulus frequency was 1 Hz and the analysis time was 50 ms.

Results

All patients experienced an immediate and substantial decrease of their pain. Assessment of the pain using a visual analogue scale (10–10) showed a decrease from 8.4 ± 1.2 to 2.6 ± 1.3. There were no complications or side-effects.

No significant changes in ipsi- and contralateral mSEPs and SCEPs were observed after cordotomy. However, the SthEPs showed a significant decrease in 10 of 12 patients (Figs. 1 and 2) provided the stimulus intensity was low (approx $0.5 - 1$ mA). With a moderately increased stimulus intensity the difference between the pre- and postlesional recordings was no longer significant.

Discussion

In an earlier study comprising 48 patients undergoing percutaneous cordotomy using CT guidance we have reported a success rate of 84% [11]. In the present study we examined somatosensory evoked potentials and found that spinothalamic tractotomy does not influence the function of the lemniscal system. However, an intraspinally located recording electrode could pick up potentials from the dorsolateral ascending fibers and there could also be a contribution from the spinothalamic tracts, were it not for the fact that in this study stimulation intensity was adjusted to the stimulation of large afferent fibers only.

CT guidance in percutaneous cordotomy has been found to be of considerable value [7,11]. Nevertheless, electric stimulation for the confirmation of a suitable placement of the tip of the lesion electrode is mandatory [6,10]. However, this requires a good cooperation with the patient who may be restless due to severe, ongoing pain; there may also be language problems. For this reason a physiological monitoring to guide the placement of the electrode without the patient's active cooperation would be helpful.

Different attempts to use various electrophysiological methods for the physiological identification of the spinal target have been made [1,2,5,8,9,10]. Most of the studies have used segmental SCEPs [1,2,8,9] but no comparisons of the responses before and after cordotomy have been made. Campbell *et al.* [2] have reported their experience with spinal cord evoked potentials in 61 patients during percutaneous cordotomy, but no attempts were made to study possible changes induced by the cordotomy lesion. Cerebral potentials evoked by direct stimulation of the thoracic spinal cord have been examined by Ertekin *et al.* [4]. They noted that relatively short latency waves could only be recorded if the tip of the stimulating electrode was located lateral to the midline indicating an activation of the dorsal columns. Taira *et al.* have also recorded cerebral responses to direct stimulation via a cordotomy electrode [8], and they described response characteristics similar to those observed in the present study. They also found that the cerebral projection of these responses differed from that of responses evoked by the activation of the lemniscal system.

Table 2. *Types of Electrophysiological Techniques, Stimulus and Recording Sites and Possible Routes of Conduction*

Mode	Terminology	Stimulus	Recording	
A	**c-mSEP** *contralateral median SEP*	median nerve contralateral to the lesion =pain side	C3'-C4' Fpz	
B	**i-mSEP** *ipsilateral median SEP*	median nerve ipsilateral to the lesion =opposite to the pain	C3'-C4' Fpz	
C	**c-SCEP** *contralateral spinal cord EP*	median nerve contralateral to the lesion	lesion electrode =spinothalamic tr.	
D	**i-SCEP** *ipsilateral spinal cord EP*	median nerve ipsilateral to the lesion	lesion electrode =spinothalamic tr.	
E	**c-SthEP** *contralateral spino-thalamic EP*	I-no motor response II-midline III-ipsilateral max. motor response lesion side = spinothalamic tr.	C3'-C4' contralateral cortex	
F	**i-SthEP** *ipsilateral spino-thalamic EP*	as in mode E lesion side = spinothalamic tr.	C3'-C4' ipsilateral cortex	
G	**SpEMyeloG** *spontaneous electromyelogram*		lesion electrode =spinothalamic tr.	

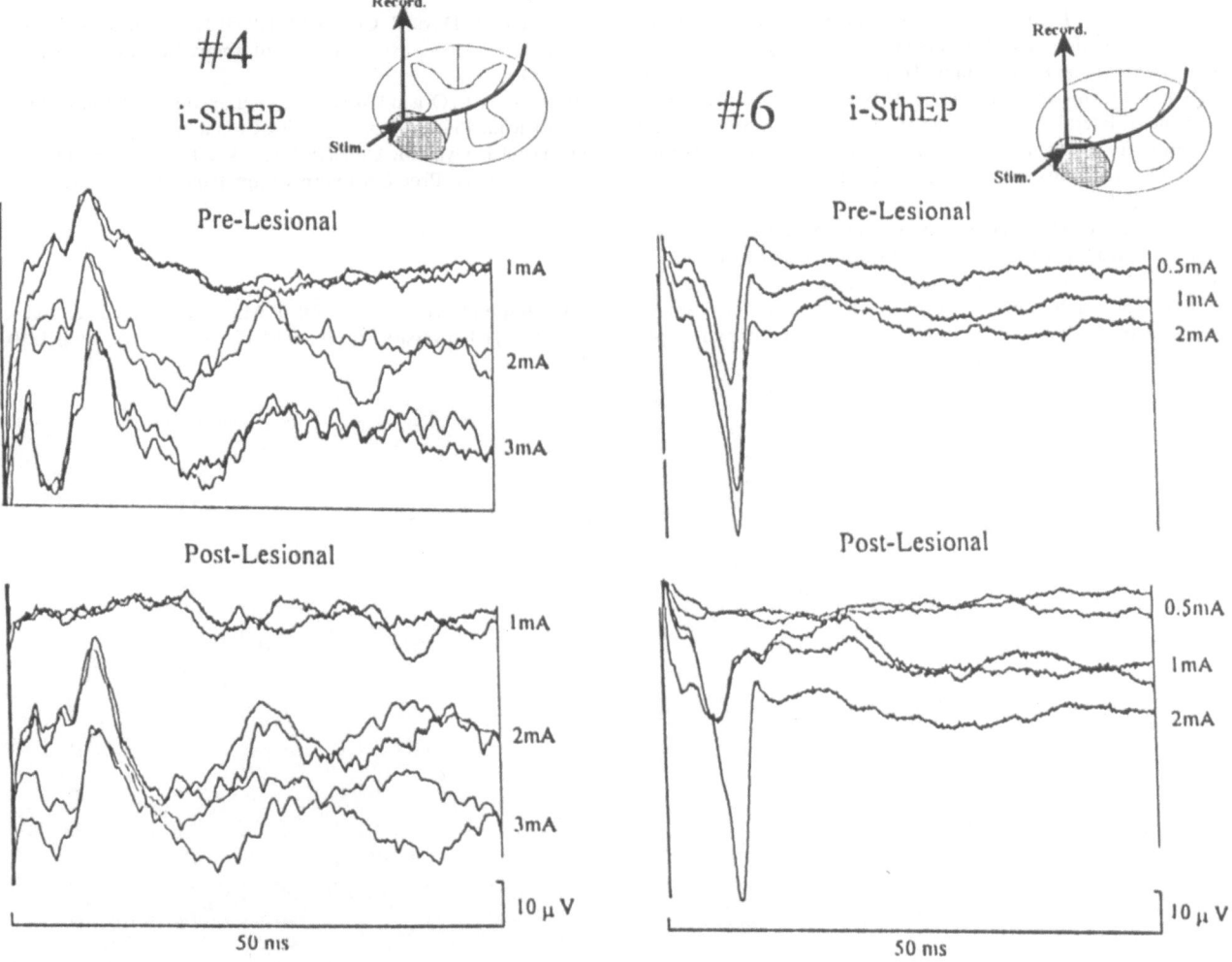

Fig. 1. StheEPs of case #4. After cordotomy (lower group of traces) a significant decrease in potentials in lower stimulation intensities was noted

Fig. 2. SthEPs of case #6. Potential depression in lower stimulation intensities after cordotomy

Cerebral responses evoked by low intensity spinothalamic stimulation was abolished by the lesion. The preservation of the response evoked by the application of higher intensity of the stimulation is presumably due to the activation of adjacent fibers outside the spinothalamic tract or of remaining, intact spinothalamic tract fibers. Since none of the patients experienced an ipsilateral motor weakness it is unlikely that the abolition of the responses after lesion was due to an injury produced in the pyramidal tract fibers. Since the pyramidal tract is located immediately posterior to the spinothalamic fibers representing the lower segments, there is always a risk that the lesion may comprise part of the pyramidal tract. It is possible that the usage of high intensity spinothalamic tract stimulation and recording of the cerebral responses could help to monitor pyramidal tract fibers when the procedure has to be done under general anaesthesia or when a satisfactory cooperation with the patient is not possible.

References

1. Amano K, Kawabatake H, Miyazaki T, Iseki H, Notani M, Kawamura H, Kitamura K (1978) Percutaneous cervical cordotomy and cerebral evoked response. Pain Abstracts, 2nd World Congress on Pain of the Int Study of Pain, Montreal, Vol 1, pp 106
2. Campbell JA, Lipton S (1984) Intraspinal evoked potentials in man during cervical cordotomy. In: Homma S, Tamaki T (eds) Fundamentals and clinical application of spinal cord monitoring. Saikon, Tokyo, pp 245–252
3. Ertekin C, Sarıca Y, Üçkardeşler L (1983) Studies on the human spontaneous electromyelogram (EMyeloG) I. Normal subjects. Electroenceph Clin Neurophysiol 55: 13–23
4. Ertekin C, Sarıca Y, Üçkardeşler L (1984) Somatosensory cerebral potentials evoked by stimulation of the lumbosacral spinal cord in normal subjects and in patients with conus medullaris and cauda equina lesions. Electroenceph Clin Neurophysiol 59: 57–66

5. Hitchcock E, Lewin M (1969) Stereotactic recording from the spinal cord of man. BMJ 4: 44–45

6. Iseki H, Amano K, Kawamura H, Tanikawa T, Kawabatake H, Notani M, Shiwaku T, Kitamura K (1982) Somatotopic arrangement of lateral spinothalamic tract in percutaneous cervical cordotomy. Proc 8th Meeting World Soc Stereotactic and Functional Neurosurgery, Part III, Zurich 1981. Appl Neurophysiol 45: 484–491

7. Kanpolat Y, Deda H, Akyar S, Bilgiç S (1989) CT-guided percutaneous cordotomy. Acta Neurochir (Wien) 46: 61–68

8. Taira T, Amano K, kawamura H, Tanikawa T, Kitamura K (1985) Cerebral-evoked responses elicited by direct stimulation of the lateral spinothalamic tract in the human. Appl Neurophysiol 48: 267–270

9. Taren JA, Davis R, Crosby EC (1969) Target physiologic correlation in stereotaxic cervical cordotomy. J Neurosurg 30: 569–584

10. Tasker RR, Organ LW (1973) Percutaneous cordotomy-physiologic identification of target site. Confin Neurol 35: 110–117

11. Yegül İ, Uyar M, Ulukaya S (1994) CT guided percutaneous cordotomy. Proc 6th Intern Congr Pain Clinics, April 15–20, pp 138

Correspondence: Mehmet Zileli, M.D., Department of Neurosurgery, Ege University Faculty of Medicine, Bornova, Izmir 35100, Turkey.

Acta Neurochir (1995) [Suppl] 64: 97–100

Transplantation of Human Chromaffin Cells for Control of Intractable Cancer Pain

Y. Lazorthes[1], J. C. Bès[2], J. Sagen[5], M. Tafani[3], J. Tkaczuk[4], B. Sallerin[1], I. Nahri[1], J. C. Verdié[1], E. Ohayon[4], C. Caratero[2], and G. D. Pappas[5]

[1]Department of Neurosurgery, [2]Department of Histology, [3]Department of Nuclear Medicine, and [4]Department of Immunology, Faculty of Medicine, University Paul Sabatier, Toulouse, France, and [5]Department of Anatomy and Cell Biology, University of Illinois, Chicago, U.S.A.

Summary

Adrenal medullary chromaffin cells produce high levels of endogenous opioid peptides. Recent data suggest that transplantation injected locally into the spinal subarachnoid space reduced intractable malignant pain. In order to determine the feasibility, the efficacy and the risks of using adrenal medullary tissue for control of irreducible pain [14, 17], we have developed a transplantation protocol on cancer pain patients selected when they required chronic intrathecal injection of morphine and progressively increasing doses to maintain the level of analgesic effects.

At the present time, our clinical trial involves 8 patients. We report here our initial results (mean follow-up: 5 months). The various data collected before and after the intrathecal administration of chromaffin cells included: 1) Pain evaluation over time, with concomitant narcotic intake, 2) CSF sampling through an implanted access port to determine the following biological parameters: biochemical assay for opioid peptides, cell count and phenotyping of lymphocytes, 3) peripheral blood samples for lymphocyte typing.

The results confirm the efficacy of adrenal medullary transplantation into spinal CSF for controlling irreducible cancer pain. Complementary intrathecal and oral morphine were totally stopped in 2 cases and stabilized in 5 others. It seems essential to have an important volume of grafted tissue to achieve analgesia with high levels of metenkephalin in CSF. A progressive decrease in met-enkephalin release was observed from 2 to 4 months after the transplantation. Two patients with a long-term follow-up (8 and 12 months) needed another intrathecal chromaffin cell graft.

Keywords: Cancer pain; opioids; neuronal graft; chromaffin cells; enkephalin.

Introduction

Different anti-nociceptive substances and notably endogenous opioid peptides and catecholamines are released from adrenal medullary chromaffin cells [10, 16]. Opioids and α-adrenergics agonists are a potent combination that synergize to produce marked analgesia [5, 15, 18].

The intrathecal allogenic transplantation of adrenal medullary tissue reduces pain sensitivity in various rodent chronic pain models: an arthritis model and a peripheral neuropathy model [10, 11]. Preliminary clinical trials have reported promising results concerning irreducible pain control in terminal cancer patients [14, 17].

The objectives of this multidisciplinary study were to: 1) Assess the feasibility, efficacy and risk of subarachnoid human adrenal medullary grafts in alleviating intractable nociceptive cancer pain, 2) Determine the place and perspectives of the method in pain control strategy in malignancy.

This protocol was approved by the Ethical Comittee (December 15, 1992) and the Hospital Review Board (C.C.P.P.R.B.) according to the French law (December 22, 1992).

Materials and Methods

The informed and consenting patients selected for the study were suffering from severe nociceptive pain of cancer origin. Intractable pain was inadequately controlled by gradual doses of slow-release oral morphine, producing intolerable side-effects. All were responding to opioids and were controlled by daily intrathecal morphine administration through an implanted injection port.

Table 1 summarizes the data concerning the first 8 patients treated by allogenic chromaffin cell transplants.

Adrenal glands were obtained from human cadaver donor tissue. Chromaffin cell grafts were prepared in our laboratory following the methodology reported earlier [14, 17]. Small pieces of adrenal medullary tissue were maintained in vitro for at least 7 days. Prior to transplantation, tissue viability was assessed using: 1) Biochemical assays in supernatant fluids (metenkephalin-Ria Kit Incstar), 2) Immunocytochemistry (monoclonal antibody to tyrosine hydroxylase- Incstar), 3) Morphological control performed with an electron microscope.

Table 1. *Clinical Features (06–93 — > 09–94)*

Patients: n = 8   Transplants: n = 10
—Cancer pain origin -rectum:3 -lung:3 -kidney:1 -uterus:1
—Nociceptive pain (lower limb ± pelvis = 5; thoracic = 3)
—Age:39–83;   mean:52 yrs (F = 1; M = 7)
  Before graft: 1. Pain score (0–4):   :   4
                 2. Karnofsky Index   :   40–80 (mean : 70)
                 3. Morphine prescription (mg/day)
                    Oral morphine   :   60–600 (mean: 260)
                    Intrathecal morphine :   2.5–30 (mean: 12)

The intrathecal lumbar graft was performed under local anesthesia by an injection into the subarachnoid space through a 14 G Touhy needle; the graft consisted of small tissue pieces suspended in the patients own cerebro-spinal fluid. Cyclosporine A had been administered (10 mg/kg/day) for 2 weeks beginning 1 day before the procedure.

Clinical and biological follow-up was carried out at day 3, day 8 and then monthly. Pain levels were determined by using a visual analogic scale (VAS) and records of daily complementary narcotic intake and functional activity (Karnofsky scale). CSF samples (10 ml) were collected through the implanted intrathecal access port to determine the following biological parameters: 1) Biochemical assay for opioid peptides, 2) Cell counts, 3) Phenotyping of lymphocytes (cytofluorometric typing for differentiation [CD4, CD8]) and activation [HLA-DR CD45] markers. Peripheral blood samples were also taken for lymphocyte typing.

Results

a) Patient Follow-up

This preliminary study involves 8 patients. The first chromaffin cell transplantation was carried out in June 1993. The mean follow-up was 148 days (from 15 to 360 days). Table 2 summarizes the clinical and biological data.

Patient no. 7 with a short follow-up (15 days) was not analyzed. Patients no. 3 and no. 5 are still alive, remain pain free and have the longest follow-up, respectively 12 and 8 months. These two patients required 2 intrathecal chromaffin cell grafts respectively at day + 40 and day + 130.

A multidisciplinary pain evaluation demonstrated a progressive decrease of pain score in 6 patients. A concomitant significant decrease in narcotics was observed in 3 cases (Table 3). The complementary analgesic drug intake was stabilized in 2 other patients (pts. no. 4 and 5) and increased in 2 cases (pts. no. 2 and 8).

Increasing levels of CSF met-enkephalin were observed in 6 patients. Table 4 summarizes the individual evolution of the met-enkephalin release in CSF.

The basic state was variable from one patient to another (from 40 to 190 pg/ml; mean: 90). The maximum increase was observed 1 month after the cell chromaffin transplant (from 80 to 580 pg/ml; mean: 303). In all the cases, we observed a secondary decrease 2 to 4 months after the transplant. A significant correlation between pain control and met-enkephalin release in CSF was noted in all cases. No side-effects were reported.

The autopsy performed at day + 15 for the terminal cancer patient no. 7 demonstrated that the grafted tissue was intact and fixed between the roots of the cauda equina.

b) Immunological Response in CSF and Blood

Three color flow: cytometric lymphocyte phenotyping was performed simultaneously in CSF and blood, for a kinetic study.

CSF: Among the 10 grafts performed, 3 patients never displayed any detectable lymphocytes; in CSF during the all monitoring period. By contrast, the majority (7/10 cases) of the patients presented lymphocytes in CSF a level higher than 1000 cells/ml at day 7 after grafting. 3 of these 7 patients had no more detectable lymphocytes after this time, whereas 4 of them exhibited persisting lymphocytes in CSF throughout the follow-up.

Table 2. *Clinical Results (n = 8)*

Patient no.	Follow-up (days)	Pain score (0–4)	Opioid intake	Met-ink basal	(pg/ml) max
1	185	0	↘ ↘	100	580
2	62	3–4	↗	40	80
3[a]	360	1	↘	80	320
4	180	2	→	40	190
5[a]	240	1–2	→	100	270
6	70	0	↘ ↘	80	210
7	15	–	→	80	80
8	60	2	↗	100	280

[a] Patients 3 and 5 required two grafts.

Table 3. *The Complementary Drug Intake Evolution (mg/24h)*

Pt. no.	Daily morphine before graft		Daily intra-thecal morphine after chromaffin cell graft												
	oral	l-Th	Day 0	+30	+60	+90	+120	+150	+180	+210	+240	+270	+300	+330	+360
1	520	10	5 / 2,5	2,5	0	0	0	0							
2	400	25	20	40	60										
3	60	2,5	2,5	2,5 (graft 2)	2,5	2,5	2,5	1	1	1	3	3	3	3	3
4	60	2,5	60 oral	0 l-Th	0	30 oral	0	60 oral	120 oral						
5	100	10	5	5	2,5	4	12 / 12 (graft 2)	12	10	10	12				
6	150	30	10	60 oral	90 oral										
7	220	10	10 + autopsy												
8	600	10	15	25	25										

Table 4. *Met-Enkephalin Release in the CSF (pg/ml)*

Pt	Day 0	+3	+8	+30	+60	+90	+120	+150	+180	+210	+240	+270	+300	+330	+360
1	100	300	320	580	280	520		250							
2		40	40	80											
3	80		130	170	280	320	150	160	130	120	90	210	230	200	190
4	40	85	150		100	150	125	190							
5	100	175	230	270	210	105	110	300	280	280	260				
6	80	150	210	190	80										
7	80														
8	100	180	280												

When lymphocytes were detectable, the CD4/CD8 ratio in CSF was relatively constant with a value ranging from 2 to 3: this data indicates a predominance of the CD4 subset of T lymphocytes in CSF. These CD4$^+$ cells were in an activated state, since CD4$^+$ CD45RO$^+$ HLA-DR$^+$ triple positive cells were twice as predominant as in CD4$^+$ lymphocytes compared to those from peripheral blood at the same time. The CD8 cells were also activated, with 40% of CD8$^+$ CD45RO$^+$ HLA-DR$^+$ in this later subset. B cells (CD19$^+$) were less than 1% of the lymphocytes in CSF, whereas in the blood this subset was at a mean of 10% of the lymphocyte population. This observation can be interpreted as a lack of the CD 19 subset in CSF after grafting.

Blood: We observed in the peripheral blood of tested patients a count of 900 cells/ml at the time of the grafting, without any significant modification during the follow-up. The mean CD4/CD8 ratio decreased from 0.98 to 0.81 two months after the graft without

any statistical significance. Two patients (4 and 5), showed signs of activation in CD8$^+$ cells (40% of CD8$^+$ CD45RO$^+$ HLA-DR$^+$ triple positive cells among the CD 8 subset).

Discussion

The earlier and larger medical use of opioids has resulted in successful pain alleviation in up to 80–90 percent of cancer pain, when the patients are managed effectively through the WHO guidelines (three-step analgesic ladder). However following prolonged use of oral opioids with increasing doses a significant percentage of patients develop tolerance phenomena and, more frequently, complications secondary to intolerable side-effects. To control these unsatisfactory results, different routes of administration have been used, especially direct intrathecal morphine administration at the spinal or cerebroventricular level [9] using implantable pumps or access ports.

The long-term efficacy of chronic intrathecal morphine has been demonstrated for nociceptive pain treatment in malignancy. Between 1978 and 1993, we have treated 240 patients by this intrathecal route, with an overall success of 84% [6–9]. The disadvantages of several daily injections include contraints in the ambulatory treatment, potential risk of infection and also pharmacological side-effects. The use of implanted pumps needs specialized costly equipment for patients with a limited life expectancy.

The objective of neural transplants, as an alternative, is to provide a long-term source of anti-nociceptive substances, reducing or eliminating the need for daily morphine administration.

This preliminary study has confirmed the initial reports [14, 17]. Transplanted human chromaffin cells can survive and continue to produce anti-nociceptive substances, acting like a "biological pump". We also have observed a significant relationship between: 1) Analgesic efficacy and CSF met-enkephalin release, and 2) Met-enkephalin levels and volume of grafted tissue. The prolonged survival of adrenal medullary transplants is perhaps limited in time, because a progressive decrease of met-enkephalin release from 2 to 4 months after the intrathecal graft has been observed. It is possible to repeat the transplantation procedure.

Even if CNS is classically considered as "immunologically privileged" [2], the appearance of lymphocytes in CSF (7/10) and in some cases (2/10) the development of peripheral activation of the immune system signify a homeostatic response against grafted cells. Long-term survival of grafted cells could be limited by the immune responses which have been observed during the analysis, or on the contrary could be favoured by an active phenomenom of lymphocyte tolerance. Allografts may ultimately be rejected [4].

In order to circumvent: 1) Shortages of human cadaver donor tissue, 2) Potential long-term rejection and 3) Necessity of repeated chromaffin cell grafts, we are investigating the possibility of using xenogeneic chromaffin cells. Analgesia has been induced by implantation of bovine chromaffin cells in rat spinal cord [12]. More recently bovine chromaffin cells immunologically isolated by encapsulaton technology has been developed [1]. Another approach is to modify the graft by removal of highly immunogenic cells so that rejection responses are circumvented by a tolerated host [3].

References

1. Aebischer P, Goddard M, Signore AP, Timpson RL (1994) Functional recovery in hemiparkinson on primates transplanted with polymer encapsulated PC 12 cells. Exp Neurol 126: 1–8
2. Barker CF, Billingham RE (1977) Immunologically privileged sites. Adv Immunol 25: 1–54
3. Cooper DKL, Kemp E, Reemtsa K, White DJG (1991) Xenoransplantation. Springer Berlin Heidelberg New York Tokyo
4. Cserr HF, Knopf PM (1992) Cervical lymphatics, the blood-brain barrier and the immunoreactivity of the brain: a new view. Immunology Today 13: 507–512
5. Drasner K, Fields HF (1988) Synergy between the antinociceptive effects of intrathecal clonidine and systemic morphine in the rat. Pain 32: 309–312
6. Lazorthes Y, Verdié JC, Bastide R, Lavados A, Descouens D (1985) Spinal versus intraventricular chronic opiate administration with implantable drug delivery devices for cancer pain. Appl Neurophysiol 48: 234–241
7. Lazorthes Y, Verdié JC, Bastide R, Clergue ML, Lavados A, Caute B, Cros J (1985) Chronic spinal administration of opiate: application in the treatment of intractable cancer pain. In: Besson JM, Lazorthes Y (eds) Spinal opioids and the relief of pain: basic mechanisms and clinical applications, Vol 127. INSERM, Paris, pp 437–463
8. Lazorthes Y, Verdié JC, Caute B, Maranhao R, Tafani M (1988) Intra-cerebro-ventricular morphinotherapy for control of chronic cancer pain. In: Fields HL, Besson JM (eds) Progress in brain research, Vol 77. Elsevier Amsterdam, pp 395–405
9. Lazorthes Y, Sallerin-Caute B, Verdié JC, Bastide R (1991) Advances in drug delivery systems and applications in neurosurgery. In: Symon *et al* (eds) Advances and technical standards in neurosurgery, Vol 18. Springer, Wien New York, pp 143–192
10. Livett G, Dean DM, Whelan LG, Lidenfriend S, Rossier J (1981) Corelease of enkephalin and catecholamines from cultured adrenal chromaffin cells. Science 225: 734–737
11. Sagen J, Pappas GD, Perlow MJ (1986) Adrenal medullary tissue transplants in the rat spinal cord reduce pain sensitivity. Brain Res 384: 189–194
12. Sagen J, Pappas GD, Pollard HB (1986) Analgesia induced by isolated bovine chromaffin cells implanted in rat spinal cord. Proc Nat Acad Sci (Washington, USA) 83: 7522–7526
13. Sagen J (1992) Chromaffin cell transplants for alleviation of chronic pain. ASSAIO Transactions 38: 24–28
14. Sagen J, Pappas GD, Winnie AP (1993) Alleviation of pain in cancer patients by adrenal medullary transplants in the spinal subarachnoid space. Cell Transplant 2: 259–266
15. Sherman SE, Loomis CW, Milne B, Cervenko FW (1988) Intrathecal oxymetazoline produces analgesia via spinal α-adrenoreceptors and potentiates spinal morphine. Eur J Pharmacol 148: 371–380
16. Wilson SP, Chang KI, Viveros OH (1982) Proportional secretion of opioid peptides and catecholamines from adrenal chromaffin cells in culture. Anesthesiology 54: 451–457
17. Winnie AP, Pappas GD, Das Gupta TK, Wang H, Ortega JD, Sagen J (1993) Subarachnoid adrenal medullary transplants for terminal cancer pain: A report of preliminary studies. Anesthesiology 79: 644–653
18. Yaksh TL, Reddy SVR (1981) Studies in the primate on the analgesic effects of intrathecal associated with intrathecal actions of opiates, alpha-adrenergic agonists and baclofen. Anesthesiology 54: 457–467

Correspondence: Y. Lazorthes, M. D., Department of Neurosurgery, CHU Rangueil, 1 av. Jean Poulhès, 31054 Toulouse Cédex, France.

Acta Neurochir (1995) [Suppl] 64: 101–105

Severe Peripheral Ischemia After Vasospasm May Be Prevented By Spinal Cord Stimulation. A Preliminary Report of a Study in a Free-Flap Animal Model

B. Linderoth[1], G. Gherardini[2], B. Ren[3], and T. Lundeberg[4]

Departments of [1]Neurosurgery and [2]Plastic Surgery Karolinska Hospital, Stockholm, Sweden, [3]Department of Neurosurgery, Central Hospital of Dalian University, Dalian, Peoples Republic of China, and [4]Department of Physiology II, Karolinska Institutet, Stockholm, Sweden

Summary

Electric spinal cord stimulation (SCS) is at present used in many centers to treat ischemic pain and ischemia in peripheral vascular disease. The most promising results have been obtained in cases where a vasospastic component is dominating. The knowledge concerning the mechanisms behind these effects has been scanty, but recent experimental studies indicate that suppression of sympathetic activity and the release of vasoactive substances may be important. A problem with many of the animals studies aimed at exploring these mechanisms is that they have almost exclusively been performed on normal animals without ischemia. However, in studies of the responsiveness of local ischemia to various pharmacological substances and to electrical transcutaneous nerve stimulation, animal models with ischemic skin flaps have been used. We applied SCS via chronically implanted electrodes in a model of local vasospasm in the rat, induced by mechanical stimulation of the vessel supplying an island flap in the groin. Male Sprague-Dawley rats were used. First, a monopolar system for spinal cord stimulation, with the intraspinal cathode at vertebral level T11, was implanted in halothane anaesthesia. After about three days of recovery the rats were anaesthetized with chloral hydrate ip and a groin neurovascular flap based on the epigastric vessels was raised. Microcirculation in the flap as well as in a control area in the contralateral groin was monitored by laser Doppler technique. Vasospasm was induced by gently pinching the superficial epigastric artery with microforceps. Two groups of animals were submitted to two spasm periods, one with SCS applied for 20 min. by 50 Hz; 0.2 msec and with 2/3 of the intensity required for a motor response before the first period. The second group, receiving sham SCS, served as a control. Both degree of ischemia after spasm provocation and the time to recovery were evaluated. In general SCS affected basal flow very little. In the control group the rats demonstrated increasing vasospastic reactions with subsequent flap ischemia to the two mechanical provocations. In the experimental group a response pattern emerged indicating that pre-spasm SCS could both reduce the spasm amplitude and significantly shorten the time for restoration of a satisfactory microcirculation in the flap. Some few trials with pharmacologically induced spasm by topical application of noradrenaline onto the feeding vessel also followed the same pattern. In conclusion, SCS seems to be able to reduce vasospasm, especially if the treatment is given before the ischemic period. This approach may supply an animal model for further studies of possible mechanisms behind the microcirculatory effects of SCS.

Keywords: Laser Doppler flowmetry; microcirculation; free flap; spinal cord stimulation; vasospasm.

Introduction

Electric spinal cord stimulation (SCS) has by now been used in many clinics to treat ischemic pain and ischemia in peripheral vascular disease for more than a decade. The best results have been obtained in cases where a vasospastic component is dominant [1, 10, 23, 28]. SCS is also effective in reducing pain in classic angina pectoris due to cardiac arteriosclerosis, as well as in patients with chest pain but with but without demonstrable vessel abnormality (syndrome X) [4]. Furthermore, studies on animal models have demonstrated that high cervical SCS may reduce the vasospasm and restore cerebral blood flow in experimental subarachnoidal hemorrhage [3, 25, 30].

Knowledge about the mechanisms behind these beneficial effects has been scanty, but recent studies indicate that suppression of sympathetic activity [17–19, 21] and the release of vasoactive substances [2, 6, 24], may be of importance.

A problem with many of the animal studies aimed at exploring the mechanisms behind the microcirculatory changes induced by SCS is that they have often been performed on normal animals without ischemia [18, 19], or on models of subarachnoid haemorrhage [3, 25, 30]. However, in studies of the responsiveness of local ischemia to various pharmacological substances and to transcutaneous electric nerve stimulation (TENS) animal models with ischemic flaps have been used [14]. Furthermore, trials with TENS on patients submitted to microsurgical procedures involving free neuro-vascular flaps with a risk of ischemia, performed by Lundeberg and Kjartansson [16, 22], demonstrated that the microcirculation in the flaps remained significantly better with electric stimulation than without.

In the present study we have applied SCS via implanted electrodes in a model of local vasospasm in the rat induced by mechanical stimulation of the vessel supplying an island flap in the groin. Previous pilot studies have indicated that SCS was much more effective if applied before spasm induction rather than after the appearance of flap ischemia. Thus we decided to test whether SCS given before spasm provocation would prevent or decrease the subsequent reduction in blood flow in the flap.

Methods

Male Sprague – Dawley rats (n = 13), were used. The study period contained two surgical sessions with a rest period of 3–5 days in between. In the first session, laminectomy at T12 was performed and a monopolar system for SCS, with the intraspinal cathode (silver; diameter 2.0 mm) at vertebral level T11 and the anode (silver; diameter 6.0 mm) in the subcutaneous tissue over the rib cage, was implanted. The leads from both electrodes were tunneled subcutaneously to a microcontact sutured to the skin in the neck of the animal. The SCS system has been described in previous reports (e.g. [20]).

After 3–5 days of recovery the rats were anaesthetized with chloral hydrate (0.4 g/kg) ip, and lying in the supine position, a groin neurovascular flap based on the epigastric vessels, was raised.

Microcirculation in the flap, as well as in a control area in the contralateral groin, was monitored by a two-channel laser Doppler system (Periflux 4001; Perimed AB, Järfälla, Sweden). Two plastic probe holders were sutured to the flap and to the contorl area before incising the skin. The preparation was allowed to stabilize for 80 minutes prior to the first spasm provocation.

Vasospasm was induced by gently champing the isolated superficial epigastric artery with two microforceps along a 5 mm length for 10 sec., as previously described by Gherardini *et al.* [7]. With this treatment the vessel was seen to contract in the operating microscope. Two groups of animals were submitted to two spasm periods: the first with SCS applied for 20 min. prior to the first provocation (Group 1; n = 6), the second (Group 2; n = 7) without any stimulation via the implanted electrodes, served as control. SCS was delivered with rectangular pulses at 50 Hz; 0.2 msec and with an intensity 2/3 of that required for a motor response (tonic contraction of the abdominal muscles), tested immediately after the beginning of anesthesia [18, 19, 21]. The SCS was generated by a Grass standard stimulator via a Grass constant current unit (Grass Instr Quincy, M, USA). The SCS intensity varied between 0.6–2.2 mA in different animals. Systemic blood pressure was recorded via a catheter in the right carotid artery.

Both LDF and blood pressure data were fed into a desk top computer with specially designed software (Perisoft Program version 4.41; Perimed AB).

The experimental set-up is illustrated in Fig. 1. One hour of stabilization was allowed after raising the flap until the start of the experiment proper. Figure 2 shows a micro-photograph of the feeding artery with the microforceps in place for spasm induction.

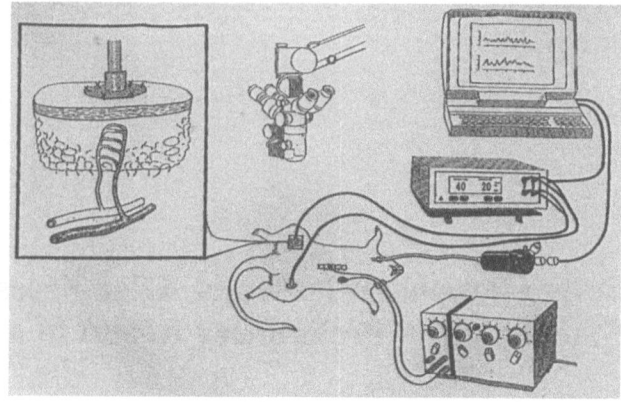

Fig. 1. The experimental set-up (description, see text). The enlarged figure on the upper left schematically pictures the flap with the feeding artery and drainage vein. The compression was selectively applied to the feeding artery

Fig. 2. The isolated superficial epigastric artery and the accompanying drainage vein supplying the island flap, as seen through the operating microscope. A pair of microforceps are ready to compress the vessel

Results

In a typical experiment an animal from the control group demonstrated profound vasospastic reactions and flap ischemia, as measured by LDF, to the two provocations, with the level of recovery deteriorating between the two spasm periods. The experimental animals with SCS prior to the first spasm period, in general recovered more rapidly and completely. Full recovery of the pre-spasm circulation was reached in

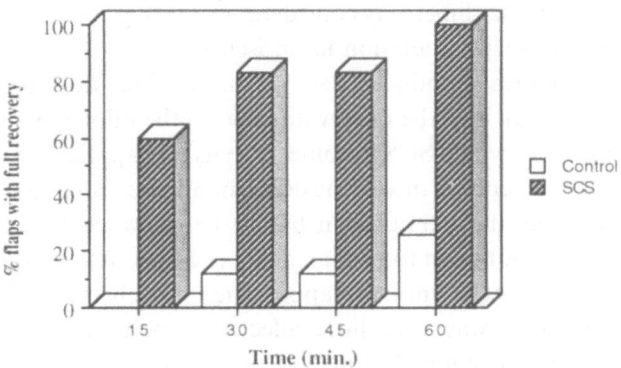

Fig. 3. Histrogram showing the percentage of flaps with full (100%) recovery of microcirculation at different times after the first spasm period

sixty minutes following the first provocation by all the animals receiving SCS before pinching, while at that time only 28% in the control group had reached this criterion. This is illustrated by Fig. 3. A 50% recovery was rapidly attained by all SCS-treated animals (in 11–12 min.). At that time only about 35% of the control animals had regained the same circulatory level. The differences between experimental and control groups were statistically significant for both these recovery criteria (P < 0.05; log rank exact test) [27].

From Table 1 it is also evident that the maximal microcirculatory flow (max. laser Doppler flux measured in arbitrary perfusion units; PU) after spasm was much higher in the SCS-group than in the control group. Furthermore, it was observed that blood flow restoration in the control flaps was almost depleted after a second spasm, while in the SCS group half of the flaps still recovered (not shown). The recordings of the systemic blood pressure did not display any significant changes during SCS, but during spasm provocation slight elevations were noted. There were no systematic differences between the two groups in this respect.

Discussion

This study demonstrates that SCS can effectively counteract one type of spasm (mechanically induced) and alleviate the resulting ischemia in the free flap rat model used here. A prerequisite seems to be that SCS is delievered before spasm induction.

Among the possible mechanisms behind decreased ischemia and increased flap survival with afferent stimulation of coarse fibers such as SCS and transcutaneously applied stimulation (TENS), the release of vasoactive substances like substance P (SP), the vasoactive intestinal polypeptide (VIP) and the calcitonin gene-related peptide (CGRP) have been discussed [2, 14, 15, 17]. Observations in favor of a role for CGRP include a much higher vasodilatory potency of this substance and the lack of tendency to evoke fluid extravasation and oedema. Furthermore SP-induced vasodilatation is dependant on an intact endothelium, while some of CGRP's capacity for vasodilatation seems to persist even after the endothelium is severely damaged, a situation present in peripheral vascular disease.

Furthermore, it has been demonstrated, both in animal experiments and in clinical studies that treatment with CGRP markedly increases the survival of circulatory compromised skin flaps [12,15]. In the latter study it was found that i.v. CGRP was even more effective than treatment with TENS. A recent study with the same flap model as used in the present study demonstrated that topical application of CGRP onto the feeding artery resulted in a faster and more complete resolution of the vasospasm than seen in untreated animals [7, 8]. Although it has been hypothe- sized that the effects of TENS on the microcirculation in these circumstances are due to a release of CGRP [22], this has not, to the best of our knowledge, been so far directly demonstrated.

Table 1. *First Spasm Induction*

Group	No.	No. (%) of flaps with complete recovery	Max post-spasm microcirculation (max. LDF flux; PU; mean +/− SEM)	Time to reach max flux (sec.; mean +/− SEM)
1. SCS	6	6 (100%)[a]	127 +/− 10[a]	2381 +/− 497
2. Control	7	2 (28%)	51 +/− 20	1982 +/− 452[a]

Different indices of the recovery of microcirculation after vasospasm induced by mechanical provocation. Number (and percentage) of flaps with complete recovery, the maximal circulation (LDF flux; PU) after a spasm period, and the time required to reach this level is indicated for the SCS and control group, respectively. [a] Indicates P < 0.05; *SEM* = standard error of the mean. [b] This average is based on the five animals with some spasm remission. The two remaining animals in this group demonstrated no recovery.

It has also been proposed that the effect of SCS in ischemia is largely due to a transient depression of sympathetically maintained peripheral vasoconstriction [17–19]. Recently it has been demonstrated that SCS acts mainly on α_1-adrenoreceptor mediated influence [21]. Since the vasospasm in e.g. Raynauds disease has been suggested as due to an increased sensitivity of the peripheral α-adrenergic receptors or to an increase in their density [5], a depression of α-receptor mediated vasoconstriction may actually constitute one component behind the effects of SCS in this condition.

Some observations further indicate that SCS is more effective if given immediately prior to a spasm period rather than one hour before it. Another group of six rats was submitted to SCS immediately before the second vasospasm in a similar experiment as that reported here (Linderoth, Gherardini, Ren and Lundeberg, unpubl. observations). This group, though displaying a rather high variability in basal levels, tended to recover better from the second spasm period than did the group in the present study with SCS given prior to the first spasm (5 flaps of 6 (83%) completely recovered to an average max flux of 138 PU, compared to 50% and an average of 116 PU in the present experimental group).

Regarding the effects of high cervical SCS on the cerebral circulation following bleeding into the subarachnoid space, only experimental data is as yet available [3, 25, 30]. Some experimental studies support an involvement of the cerebral sympathetic innervation and the release of prostaglandins (e.g. [6]). There are data indicating that also in this territory a decrease of sympathetic activity may be a critical factor in the effect of SCS [29]. Recent data from this group, in parallell to our observations in the present study, demonstrate that SCS may have a preventive effect on the early vasospasm following experimental subarachnoid haemorrhage in the rabbit [30].

In some pilot experiments using the same model as in this study, topical application of noradrenaline (gelfoam soaked in 2% noradrenaline/saline solution) onto the feeding artery induced vasospasm similar to that obtained with mechanical provocation. SCS given before application seemed able to diminish also this type of vasospasm. Further studies utilizing pharmacologically induced vasospasm are in progress.

Conclusions

The observations in this study demonstrate that SCS, applied with current parameters similar to those used in the clinic, may considerably increase the recovery of microcirculation in an ischemic skin flap after mechanically induced vasospasm. Taking into consideration also the observations from the pilot experiments in which SCS in some animals was applied only after the appearance of the deficient microcirculation, it seems that stimulation has to be delivered before spasm induction to be maximally effective. When SCS was withheld until the appearance of ischemia, this condition was very little affected compared to the situation in control animals. This observation is also in accordance with clinical reports from patients treated by SCS for Raynauds disease, where a much better effect from stimulation is obtained if used before provocation of ischemic pain, e.g. by walking in cold weather, than if SCS is given when the pain and the pallor have appeared.

Although the underlying mechanisms remain obscure, it seems that depression of sympathetic activity and the release of vasoactive substances locally may be of importance. Our approach may supply a model for further studies of possible mechanisms behind the microcirculatory effects of SCS in ischemia.

Acknowledgements

The skilful assistance of Göte Hammarström in manufacturing the SCS electrodes is gratefully acknowledged. The wires were generously supplied by Medtronic Inc., Minn., USA. This project was supported by grants from Karolinska Institutet, Svenska Läkaresällskapet and from Svenska Sällskapet för Medicinsk Forskning.

References

1. Arregui R, Morandeira JR, Martinez G, Gomez A, Calatayud V (1989) Epidural neurostimulation in the treatment of frostbite. PACE 12(2): 713–717

2. Broseta J, Sanchez-Ledesma MJ, Goncalves J (1994) Humoral mechanisms mediating in the increase of peripheral blood flow following spinal cord stimulation: An experimental study in the dog. In: Herreros J et al (eds) Spinal cord stimulation for peripheral vascular disease. Advances and controversies. Editorial Libro del Ano, SL, Madrid, 17–23

3. Broseta J, Sanchez-Ledesma MJ, Garcia-March G (1994) Effect of high cervical spinal cord stimulation on cerebral blood flow in experimental vasospasm. In: Herreros J et al (eds) Spinal cord stimulation for peripheral vascular disease. Advances and controversies. Editorial Libro del Ano, SL, Madrid, pp 209–214

4. Eliasson T, Albertsson P, Hårdhammar P, Emanuelsson H, Augustinsson L-E, Mahnnheimer C (1993) Spinal cord stimulation in angina pectoris with normal coronary arteriograms. Coronary Artery Dis 4: 819–927

5. Freedman RR, Sabharwal SC, Desai N, Wenig P, Mayes M (1989) Increased α-adrenergic responsiveness in idiopatic Raynaud's disease. Arthritis Rheum 32: 61–65

6. Garcia-March G, Sánchez-Ledesma MJ, Anaya J, Broseta J (1989) Cerebral and carotide haemodynamic changes following

cervical spinal cord stimulation. An experimental study. Acta Neurochir (Wien) [Suppl] 46: 102–104

7. Gheradini G, Jernbeck J, Samuelson U, Heden P (1995) Effects of calcitonin gene-related peptide and lidocaine on mechanically induced vasospasm in an island flap in the rat. J Reconstruct Microsurg 11: 179–183

8. Gheradini G, Lundeberg T, Gazelius B, Brodda-Jansen G, Samuelson U (1995) Calcitonin gene-related peptide increases microcirculation after mechanically induced ischemia in an experimental island flap. Plast Reconstr Surg, in press

9. Gil-Salú JL, Gonzalez-Darder JM (1992) Changes in mechanical responsiveness of the rat aorta after cervical spinal cord stimulation. Acta Neurochir (Wien) 117: 206–209

10. Ijima H, Koishizawa T, Miya J, Mitsui T, Hori M (1987) Spastic arterial disease of the lower extremity. 2nd Int PVD Symposium, Cyprus, Abstract, p 47

11. Jernbeck J (1992) Calcitonin gene-related peptide (CGRP) as a vasodilator in man. Morphological, physiological and clinical aspects. Thesis. Karolinska Institutet, Stockholm, 47 pp

12. Jernbeck J, Dalsgaard C-J (1993) Calcitonin gene-related peptide (CGRP) treatment of flaps with compromised circulation in man. Plast Reconstr Surg 91(2): 236–242

13. Jurell G, Jonsson C-E (1976) Increased survival of experimental skin flaps in rats following treatment with antiadrenergic drugs. Scand J Plast Reconstr Surg 10: 169–172

14. Kjartansson J (1988): Peripheral sensory neurons and viability of skin flaps. Thesis. Karolinska Institutet, Stockholm, 30 pp

15. Kjartansson J, Dalsgaard C-J (1987) Calcitonin gene-related peptide increases survival of a musculocutaneous critical flap in the rat. Eur J Pharmacol 142: 355–358

16. Kjartansson J, Lundeberg T (1990) Effects of electrical nerve stimulation (ENS) in ischemic tissue. Scand J Plast Reconstr Hand Surg 24: 129–134

17. Linderoth B (1992) Dorsal column stimulation and pain: experimental studies of putative neurochemical and neurophysiological mechhanisms. Thesis. Karolinska Institutet, Stockholm, 67 pp

18. Linderoth B, Fedorcsak I, Meyerson BA (1991) Peripheral vasodilatation after spinal cord stimulation: animal studies of putative effector mechanisms. Neurosurgery 28: 187–195

19. Linderoth B, Gunasekera L, Meyerson B (1991) Effects of sympathectomy on skin and muscle microcirculation during dorsal column stimulation: animal studies. Neurosurgery 29: 874–879

20. Linderoth B, Stiller C-O, ÓConnor WT, Hammarström G, Ungerstedt U, Brodin E (1993) An animal model for the study of brain

transmittor release in response to spinal cord stimulation in the awake freely moving rat: preliminary results from the PAG. In: Meyerson BA et al (eds) Advances in stereotactic and functional neurosurgery, Vol 10. Acta Neurochir (Wien) [Suppl] 58: 156–160

21. Linderoth B, Herregodts P, Meyerson B (1994) Sympathetic mediation of peripheral vasodilatation induced by spinal cord stimulation: animal studies of the role of cholinergic and adrenergic receptor subtypes. Neurosurgery 35(4): 711–719

22. Lundeberg T, Kjartansson J, Samuelsson UE (1988) Effect of electrical nerve stimulation on healing of ischaemic skin flaps. Lancet 24: 712–714

23. Robaina FJ, Dominguez M, Diaz M, Rodriguez JL, de Vera JA (1989) Spinal cord stimulation for relief of chronic pain in vasospastic disorders of the upper limbs. Neurosurgery 24: 63–67

24. Sanchez-Ledesma MJ, Garcia-March G, Goncalves J, Ananya J, Silva I, Gonzalez-Buitrago JM, Navajo JA, Broseta J (1990) Role of vasoactive substances in the segmentary vasomotor response following spinal cord stimulation. Stereotact Funct Neurosurg 54: 224–231

25. Sanchez-Ledesma MJ, Garcia-March G, Silva I, Robaina F, Broseta J (1991) Effect of cervical spinal cord stimulation on experimental cerebral vasospasm in the rat. 9th European Congress of Neurosurgery, Moscow, Abstract, p 304

26. Seaber AV (1987) Experimental vasospasm. Microsurgery 8: 234–241

27. Siegel S, Castellan NJ Jr (1988) Nonparametric statistics for the behavioral science, 2nd Ed. McGraw-Hill, New York, 399 pp

28. Vignotto V, Pulatti P, Baso AM, Maritano M (1987) Spinal electrostimulation and Raynaud's syndrome. Preliminary results. 2nd Int PVD Symposium, Cyprus, Abstract p 113

29. Visocchi M, Cioni B, Meglio M, Puca A, Vergari A, Argiolas L (1993) Modulation of cerebrovascular sympathetic tone during spinal cord stimulation: an experimental study. In: Galley D, Illis LS, Krainick JU, Meglio M, Sier JC, Staal MJ (eds) Proceedings of the 1st Congr of the Internat Neuromod Soc, Rome, 1992. Monduzzi, Bologna, pp 59–63

30. Visocchi M, Marano G, Cioni B, Meglio M (1994) Protective effect of SCS on early cerebral vasospasm in the rabbit. Preliminary results. 2nd Congr of the Internat Neuromod Soc, Gothenburg, Abstract, p 7

Correspondence: Bengt Linderoth, M.D., Dr. med. Sci., Department of Neurosurgery, Karolinska Hospital, S-17176 Stockholm, Sweden.

Acta Neurochir (1995) [Suppl] 64: 106–108
© Springer-Verlag (1995)

Spinal Cord Stimulation Versus Reoperation for Failed Back Surgery Syndrome: a Prospective, Randomized Study Design

R. B. North, D. H. Kidd, and S. Piantadosi

Departments of Neurosurgery and Biostatistics, The Johns Hopkins University School of Medicine, Baltimore, MD, U.S.A.

Summary

Retrospectively reported results of spinal cord stimulation compare favorably with those of neurosurgical treatment alternatives for the treatment of failed back surgery syndrome, including reoperation and ablative procedures. There has been no direct prospective comparison, however, between SCS and other techniques for pain management.

Therefore, we have designed a prospective, randomized comparison of spinal cord stimulation and reoperation in patients with persistent radicular pain, with and without low back pain, after lumbosacral spine surgery. Patients selected for reoperation by standard criteria are randomly assigned to initial treatment by one or the other technique. The primary outcome measure is the frequency of crossover to the alternative procedure, if the results of the first have been unsatisfactory after 6 months. Results for the first 27 patients reaching the 6-month crossover point show a statistically significant ($p = 0.018$) advantage for spinal cord stimulation over reoperation. Many other potentially important outcome measures will now be followed long-term as a larger overall study population accumulates.

Keywords: Spinal cord stimulation (SCS); reoperation; failed back surgery syndrome; pain.

Introduction

Each year more than 200,000 patients in the United States undergo lumbosacral spine surgery; of these, between 20 and 40% experience persistent or recurrent pain [10]. Consequently, "failed back surgery syndrome," in which pain persists or recurs after surgery, must be considered a common condition. Yet after decades of multidisciplinary treatment and clinical research, failed back surgery syndrome in many patients is refractory to medical, surgical, and behavioral therapy.

In a number of retrospective case series, spinal cord stimulation (SCS) has been reported to be effective for failed back surgery syndrome. Indeed, among the most common neurosurgical treatments for the syndrome (reoperation for decompression and/or stabilization, ablative procedures such as rhizotomies, denervations, and ganglionectomies, and spinal cord stimulation), disinterested third-party long-term (5-year) follow-up has indicated that spinal cord stimulation has a substantially higher rate of success [6–8]. This is of interest, because the technique is less invasive, and has lower morbidity than do reoperation or ganglionectomy. Moreover, patients' abilities to perform everyday activities and their neurologic function have been enhanced. Even though patient selection criteria for these procedures have not been uniform, and hence direct comparisons among these series are somewhat questionable, it is clear from the most obvious differences that patients chosen for spinal cord stimulation should be among the most difficult to treat: They have had symptoms for the longest period of time, with the most severe radiographic and clinical evidence of neurologic disease (inoperable arachnoid fibrosis in many cases), and they have undergone more prior, unsuccessful procedures than have patients chosen for reoperation.

As suggested by the results of these retrospective series, we have designed a prospective, randomized study of SCS and reoperation, in patients who meet standard criteria for both procedures. In contrast to prior studies in which we reserved SCS as a procedure of last resort after reoperation, the present study addresses SCS as an alternative to reoperation, and as "late" rather than "last" resort.

Methods

Study Population

Study candidates are selected from a large referral population with failed back surgery syndrome. Patients are included if they have 1) surgically remediable disease, which is 2) competent to explain their complaints of radicular pain, with or without low back pain. Standard clinical and radiographic criteria for surgical intervention for lumbosacral spine disease are observed [1,5]. 3) A confirmatory second opinion by a neurosurgeon or spine surgeon is obtained in all study patients before reoperation. Patients are excluded from the study for any of the following:

(1) A major or disabling neurologic deficit (e.g. foot drop, neurogenic bladder), in the distribution of a nerve root or roots with surgically remediable compression. Such patients undergo reoperation.
(2) Radiographically critical neural compression (e.g. extremely large disc fragment, or severe central stenosis, with myelographic block or its CT/MRT equivalent). Again, such patients undergo reoperation.
(3) Radiographic evidence of gross instability requiring fusion.
(4) Significant untreated dependency on prescription narcotic analgesics or benzobiazepenes.
(5) Major psychiatric comorbidity evident clinically or on routine psychological testing.
(6) The presence of any other clinically significant or disabling chronic pain problem.
(7) A chief complaint of axial (low back) pain, exceeding radicular pain – i.e., buttock and leg pain.

Study Design

All patients undergo baseline, standardized psychological testing, and a baseline, quantitative evaluation of functional capacities by a physical therapist, before randomization. Those randomized for operation then undergo surgery by one or more of several participating surgeons. Patients randomized to spinal cord stimulation are treated by percutaneous placement of a temporary electrode (3487A Pisces Quad, Medtronic, Inc., Minneapolis, MN), for a routine 2-1/2 day trial. If a patient reports at least 50% estimated relief of pain, while demonstrating stable or improved medication intake, and improved physical activity commensurate with neurologic status and age, a permanent implant (3487A-56, 3470 Xtrel) is offered. If these criteria are not met, the patient is offered early crossover to reoperation. This provides the primary outcome measure: that is, the frequency of crossover from one treatment, which has thereby "failed," to the other.

Patients are contacted by a disinterested third party, who has not been involved in their treatment, for assessment of outcome 6 months after the initial procedure, by use of a standardized questionnaire. The same individual reminds the patient of the option of crossing over to the alternative procedure, if the results of the first procedure have not been satisfactory. (In routine follow-up by the surgeon, this subject is not broached, and is addressed only in response to a specific request or complaint by the patient.) Six-month follow-up evaluation includes diagnostic imaging studies for reoperation patients, to ensure that the goals of surgery have been achieved; repeat functional capacity assessment by physical therapists; and follow-up psychological testing.

Results

Of the first 81 patients found to be eligible for the study, 51 have consented to randomization. The remaining 30 opted for reoperation, outside the study, but reserved the option of SCS if reoperation was unsuccessful. All these patients have been followed, however, to allow inferences with respect to the external validity of the study (i.e. generalization to the entire group, beyond those consenting to randomization).

Twenty-seven of the randomized patients have reached the 6-month follow-up point, at which time they became eligible for crossover. The following initial results were observed: of 15 patients undergoing reoperation, 10 (67%) opted for crossover to spinal cord stimulation: 2 of 12 (17%) of the patients undergoing spinal cord stimulation initially, opted for crossover to reoperation (p = 0.018, Fischer's exact test, two-tailed). Outside the study, 8 of 19 patients (42%) who reached 6-month follow-up after reoperation opted for "crossover" to SCS.

Although they may not have failed the primary outcome measure, there are patients in both groups who clearly are not treatment "successes." In particular, two of our SCS patients obtained only limited relief from the temporary electrode, and have not yet elected to proceed with the permanent implant; nevertheless, they have not opted for reoperation. These and other patients with ambiguous treatment results continue with routine physical therapy; a number of other outcome measures are to be assessed.

Discussion

Our primary outcome measure (the frequency of crossover from one treatment, which has thereby "failed," to another) showed a statistically significant advantage of SCS over reoperation for the failed back surgery syndrome for the first half of our planned randomized study population.

This is an interim evaluation, rather than a preplanned assessment for early termination of the study; the latter has special statistical implications. Moreover, although the primary outcome measure of this study is very straightforward, and is germane to issues of health care utilization, outcome remains to be assessed by a number of secondary measures. The study continues to accrue patients, for a planned sample size of fifty patients. At that time, several other outcome measures will be studied: ratings of pain and its relief; medication use; work status and activities of daily living; functional capacity as measured quantitatively by our physical therapists; and psychological test results. Further, data on health care costs are collected for a series of 40 of the 50 randomized patients.

There are many treatment options for the failed back surgery syndrome. Neurosurgical treatment options include (1) anatomic procedures – attempting to address the structural cause of pain – i.e., reoperation; (2) augmentative procedures, such as spinal cord stimulation; and (3) ablative procedures. In our experience, major ablative procedures for FBSS, such as dorsal root ganglionectomy, have not proven successful [6], and minor ablative procedures such as radiofrequency facet neurotomy have a limited role [9]. Therefore, in this study, we have addressed the anatomic and augmentative neurosurgical treatment options for FBSS. We also standardized other non-surgical treatmens, in particular rehabilitation and physical therapy, across patients in this study. Although such treatments may be satisfactory therapy in certain of these patients [2]; the present study cannot address this question.

We have excluded patients with a chief complaint of axial low back pain from this study, because they are difficult to treat with the usual, contemporary SCS devices; achieving overlap of the low back by stimulation paresthesias is technically difficult. There are now available specialized electrode geometries and testing methods for this patient population [3, 4]; but these were beyond the scope of this study.

In conclusion, the interim results of this study, which is the first prospective, randomized comparison of SCS with any other treatment for pain, indicate that the role of spinal cord stimulation can be expanded, as an alternative to reoperation.

Acknowledgement

Support for this study has been provided by Medtronic, Inc., Minneapolis, MN. The sponsor has had no involvement with any aspect of study design, conduct or analysis. As per Johns Hopkins University guidelines, none of the authors or study personnel have any equity interest in any manufacturer of spinal cord stimulation devices.

References

1. American Association of Neurological Surgeons(1989) Neurosurgical Case Screening Guidelines. AANS, Park Ridge, Illinois
2. Cassisi JE, Sypert GW, Salamon A, Kapel L (1989) Independent evaluation of a multidisciplinary rehabilitation program for chronic low back pain. Neurosurgery 25: 877–883
3. Law JD (1987) Targeting a spinal stimulator to treat the "failed back surgery syndrome". Appl Neurophysiol 50: 437–438
4. Law JD, Kirkpatrick AF (1991) Pain management update: spinal cord stimulation. Am J Pain Management 2: 34–42
5. Long DM, Filtzer DL, BenDebba M, Hendler NH (1988) Clinical features of the failed-back syndrome. J Neurosurg 69: 61–71
6. North RB, Kidd DH, Cambell JN, Long DM (1991) Dorsal root ganglionectomy for failed back surgery syndrome: a five year followup study. J Neurosurg 74: 236–242
7. North RB, Cambell JN, James CS, Conover-Walker MK, Wang H, Piantadosi S, Rybock JD, Long DM (1991) Failed back surgery syndrome: Five-year follow-up in 102 patients undergoing reoperation. Neurosurgery 28: 685–691
8. North RB, Ewend MG, Lawton MT, Kidd DH, Piantadosi S (1991) Failed back surgery syndrome: five-year follow-up after spinal cord stimulator implantation. Neurosurgery 28: 692–699
9. North RB, Han M, Zahurak M, Kidd DH (1994) Radiofrequency lumbar facet denervation: analysis of prognostic factors. Pain 57: 77–83
10. Wilkinson HA (1991) The failed back syndrome: etiology and therapy, 2nd Ed. Harper and Row, Philadelphia

Correspondence: R. B. North, M. D., Department of Neurosurgery, The Johns Hopkins University School of Medicine, 600 North Wolfe Street/Meyer 7–113, Baltimore, MD 21287–7713, U.S.A.

Acta Neurochir (1995) [Suppl] 64: 109–115
© Springer-Verlag (1995)

Spinal Cord Stimulation Versus Spinal Infusion for Low Back and Leg Pain

S. J. Hassenbusch[1], **M. Stanton-Hicks**[2], and **E. C. Covington**[2]

[1]Department of Neurosurgery, M.D. Anderson Cancer Center Houston, Texas, U.S.A. and [2]Pain Management Center, Cleveland Clinic Foundation Cleveland, Ohio, U.S.A.

Summary

The relative roles of spinal cord stimulation and the spinal infusion of opioids in the treatment of chronic, non-cancer lower body pain remains unclear. This report contains a retrospective analysis of patients with chronic lower body, neuropathic pain and treated over a 5 year period. Unilateral leg and/or buttock pain was treated initially with spinal stimulation and bilateral leg or mainly low back pain was treated initially with spinal infusions. 26 patients received spinal stimulation. Pain relief was ≥ 50% in 16 (62%) with increased activity levels. Stimulator coverage was most difficult or failed in patients with buttock pain. 16 patients received long-term spinal infusions. Pain relief was ≥ 50% in 2 (13%) but 25–49% in another 8 (50%) with stable infusion doses and was best in patients requiring low-dose (< 1 mg/h morphine intrathecal) infusions in the trial period. The review indicates that spinal infusions may be best for bilateral or axial pain that has not responded to spinal stimulation. Clonidine appears to be an alternative in high-dose morphine patients. New diamond-shaped electrode and dual quadripolar arrays appear to be very helpful for back, buttock, and/or bilateral leg pain patterns.

Keywords: Spinal cord stimulation; pump; pain; morphine; spinal opioids; clonidine.

Introduction

The treatment of chronic non-cancer lower body pain remains one of the more difficult challenges in medicine. For those patients with intractable pain that has not responded to various medications, physical therapy treatments, methods of immobilization, and nerve blocks, spinal cord stimulation and the long-term spinal infusion of opioids remain effective modalities. Although both have been used as treatments for a number of decades, the relative value and efficacy of these two treatments remains unclear. This paper presents a review of the experience at one institution with each of these modalities in the treatment of intractable, non-cancer lower body pain.

Methods and Materials

Eligibility for Study

Patients with chronic lower body pain, predominately neuropathic in character, were included in this study over a five-year period. The pain must have been either midline lower back pain and/or unilateral or bilateral leg pain. All of the patients were thoroughly assessed by a multidisciplinary group involving anesthesiology, neurosurgery, psychiatry, rehabilitation medicine, nursing, orthopedics, and other specialties as needed for individual patient care. The study is a retrospective analysis of patients who were treated uniformly by these authors at the Cleveland Clinic Foundation over the period 1988 to 1993. Pain severity was rated using a verbal digital pain scale (VDS) with the following question "On a scale of zero to ten where zero is no pain and ten is the worst pain you could ever imagine, what is your pain now?"[7].

The intractable nature of the pain was determined by trials of various non-opioid analgesics and other adjuvant medications as well as opioid medications given by oral routes. All patients were assessed for any evidence of spinal instability. Spinal stabilization and/or back immobilization using back bracing were utilized as appropriate. Various blocks and/or denervations were also used, as appropriate, to provide pain relief for the patients.

The patients in this series had pain rated at a severity of six or greater, limitation of functional activity because of the pain, no contraindications to the placement of a permanent stimulator or spinal infusion pump, and pain with a predominately neuropathic character involving either the midline lower back and/or one or both legs. Neuropathic pain characteristics were based upon the mechanism of causation for the pain, including some evidence of nerve injury, and the pain character with descriptions such as shock-like, shooting, lancinating, or electrical, possibly with a burning or dysesthetic component [19].

Testing and Implantation of Systems

For patients with radicular pain involving one leg with or without unilateral buttock pain, a trial of spinal cord stimulation was recommended first. For patients with midline back pain and/or bilateral leg pain, a trial of the long-term spinal infusion of opioid was recommended first. If the patients failed screening with either of these modalities, the other was then tested. This treatment strategy was based upon the available stimulator hardware during this time period which provided some limitation in producing stable stimulation patterns for bilateral or midline pain. It was also based upon the relatively new use of the long-term intrathecal infusion of opioid in the treatment of non-cancer lower body pain.

For trial of a spinal cord stimulator, a percutaneous catheter-type electrode (Pisces electrode, Medtronic Corporation, Minneapolis, MN) was used. The quadripolar arrangement was then tested over 3–5 days. During this testing period, pain relief and improvements in activity levels were assessed. If this reduced the patient's pain by ≥ 50%, the system was internalized and connected to an implanted subcutaneous generator (Itrel II, Medtronic Corporation, Minneapolis, MN).

For testing of the possible long-term intrathecal infusion of opioid, a percutaneous temporary intrathecal catheter was placed and connected to an external infusion pump. Through this was infused morphine sulfate starting at a dose of 0.1 mg/h and increasing to a maximum dose of 1.5 mg/h. If the patient showed intolerable side effects, such as nausea, vomiting, or altered mental status, sufentanil citrate was infused to a maximum dose rate of 1.5 μg/h. If this provided pain relief and increased activity as described above for spinal cord stimulation, a permanent system was implanted. This system consisted of a silicon elastomer intrathecal catheter and an implanted programmable infusion pump (Synchromed pump, Medtronic Corporation, Minneapolis, MN).

Follow-up Evaluations

After stimulator or subcutaneous infusion pump placement, the patients were seen every 2–4 weeks. At each follow-up evaluation, the devices were reprogrammed to provide optimal pain relief. The pump was refilled as needed and consideration was given to a change from morphine to sufentanil or vice-versa if long-term pain relief was < 25% despite maximal doses. The maximum infusion rate for morphine was 3 mg/h and for sufentanil 3 μg/h. At each follow-up evaluation, the patient's pain was rated, using a VDS scale, and activity levels determined. Any side effects or complications were also ascertained.

Results

Description of Patients

Over the five-year period, 42 patients meeting these criteria underwent the implantation of either a spinal cord stimulator or a spinal infusion pump. For spinal cord stimulation, five patients were screened without adequate pain relief. Three of these patients received trial spinal infusions and found effective pain relief during the screening phase. The other two patients did not undergo any other testing. For spinal infusion of opioid on a long-term basis, four patients underwent trialing but did not achieve adequate pain relief. Spinal cord stimulation was not tested in any of these four patients because of concerns that an adequate pattern of stimulation could not be obtained due to the complex pattern of pain which bilateral lower back and both legs in all of the four patients.

Spinal Cord Stimulation

Greater than 50% long-term pain relief was obtained in 16 of the 26 patients who underwent place-

ment of a sinal cord stimulator (Table 1). Four patients had 25–49% relief and the remaining 6 patients had minimal or no long-term relief. Although initial screening was good in these 6 failure patients, the pain relief decreased over the period from 3–19 months after stimulator placement. For the successful patients, activity levels increased by 9.3% ± 2.3% and three patients returned to work or increased to full-time employment.

The biggest problem with the stimulator patients was in achieving and maintaining a stimulation pattern that matched the area of pain. Of the 26 patients, one underwent a revision from a catheter-type electrode to a plate-type quadripolar electrode (Resume Electrode, Medtronic Corporation, Minneapolis, MN). Fainures in many of these spinal cord stimulation patients were related to areas of pain that increased to become either bilateral or midline. Difficulty in obtaining a bilateral or a midline area of stimulation was the most common reason for long-term failure. Two of these patients had an epidural area of approximately 3 cm diameter that would provide excellent stimulation. Problems in these patients, however, occurred in maintaining an electrode position over this very small area on a long-term basis.

The mean follow-up time was 2.1 ± 0.3 years. Three patients used Schedule III opioid medications on an infrequent (1–2 tablets every several days) basis. None of the other patients used any narcotic medications at the last evaluation. There were no technical problems left with the stimulator hardware, including infections, disconnections, or generator failure. Five patients did require repositioning of catheter-type electrodes and two patients revision of the stimulator generator.

Long-Term Intrathecal Infusion of Opioid

Sixteen patients underwent the permanent placement of a programmable infusion pump connected to an intrathecal catheter (Table 2). Ten of these patients had long-term pain relief and six had failed to obtain continued pain relief after 10 months to 3.5 years. The average length of follow-up for all 16 patients was 2.0 ± 0.3 years. In the successful patients, mean pain reduction by VDS scores was 41.8% ± 3.6% with activity improvement of 10.3% ± 3.6%. Of four patients working preimplantation, two eventually stopped working. None of the patients unemployed preimplantation returned to work afterwards.

The dose rates in these patients appeared to fall into two different groupings. Approximately half of the

Table 1. Patients Receiving Spinal Cord Stimulation

ID	Sex	Diagnosis	VDS[a] pre-op	Activity pre-op	Age	Date implant	Type system[c]	Later op and time since implant	Later op and time since implant	Later op and time since implant	Length follow-up (yrs)[d]	VDS[a] last follow-up	Activity last follow-up[b]	Assessment pain relief
1	F	arachnoiditis	9	80	41	09/15/88	catheter	change to plate electrode 3 mos			0.4	9	70	none
2	F	epidural scar	8.5	80	74	10/03/88	catheter				6.0	4	80	good
3	F	arachnoiditis	9	60	65	10/27/88	catheter				0.7	9	60	none
4	M	arachnoiditis	9	90	42	01/19/89	catheter	revise electrode 10 mos	revise electrode 1.5 yrs	revise electrode 4.3 yrs	5.5	3.5	100	good
5	F	CRPS (RSD)[e]	9.5	60	36	11/02/90	plate	revise electrode 0.7 yrs	revise electrode 1.6 yrs	revise generator 2.5 yrs	4.1	5	70	fair
6	J	intercostal neuropathy	8	70	30	10/22/91	plate	revise electrode 0.3 yrs	revise electrode 0.4 yrs	revise electrode 0.9 yrs	0.9	8	60	none
7	F	peripheral neuropathy	9	60	65	12/03/91	plate				2.1	4	70	good
8	F	intercostal neuropathy	9.5	80	28	12/11/91	plate	revise generator 0.3 yrs			1.6	8	80	none
9	F	arachnoiditis	8	70	34	2/20/92	catheter				1.9	3	80	good
10	M	epidural scar	9	90	38	02/20/92	plate				6.0	3.5	100	good
11	M	RSD	10	60	29	03/24/92	catheter	revise electrode 0.5 yrs			0.4	9	60	none
12	M	spinal cord injury	9	80	44	03/24/92	plate				1.9	2	80	good
13	M	CRPS (RSD)	10	60	41	03/31/92	plate				2.7	4	70	good
14	F	CRPS (RSD)	10	60	32	05/15/92	catheter				2.1	6.5	70	fair
15	F	peripheral nerve injury	8	70	34	06/16/92	catheter	revise generator 0.1 yrs			0.3	8.5	60	none
16	F	CRPS (RSD)	9	80	31	09/15/92	catheter	revise electrode 0.2 yrs	revise electrode 0.3 yrs		2.0	1.5	80	good
17	F	arachnoiditis	9.5	70	68	09/15/92	plate				2.3	5.5	80	fair
18	F	arachnoiditis	8.5	60	66	10/13/92	catheter				1.9	2.5	70	good
19	F	CRPS (RSD)	10	50	27	12/8/92	catheter				1.5	3	70	good
20	F	CRPS (RSD)	9	80	35	02/02/93	catheter				1.5	4	80	good
21	M	spinal cord injury	7	70	36	02/09/93	plate				1.3	4.5	70	fair
22	F	CRPS (RSD)	8	80	50	02/23/93	catheter				1.3	2	90	good
23	F	arachnoiditis	7.5	80	42	04/06/93	catheter				1.6	3.5	80	good
24	F	CRPS (RSD)	8.5	70	35	05/25/93	catheter				1.7	1.5	70	good
25	F	arachnoiditis	9	90	71	07/13/93	catheter				1.2	3	90	good
26	M	arachnoiditis	8.5	80	45	07/16/93	catheter				1.0	2	80	good

[a] VDS verbal digital scale. [b] Activity rating: activity rating: 100 = job, full-time, 90 = job, part time, 80 = drives a car ≥ 1x/month, 70 = out of house and property ≥ 2x/month, 60 = out of house and property <2x/month, 50 = out of house but not-off property, 40 = does household chores, 30 = no household chores but out of bed ≥ 6 hrs/day, 20 = out of bed < 6 hrs/day, 10 = bedbound, 0 = dead. [c] Catheter catheter-type epidural electrode, plate plate-type epidural electrode. [d] Length of time from implant to last follow-up or classification as failure of longterm pain relief. [e] CRPS complex regional pain syndrome which is new terminology for reflex sympathetic dystrophy (RSD).

Table 2. *Patients Receiving Spinal Infusions*

ID	Sex	Diagnosis	VDS[a] pre-op	Activity pre-op[b]	Age	Date implant	Drug infused	Later op and time since implant	Later op and time since implant	Later op and time since implant	Length follow-up (yrs)[c]	VDS[a] last follow-up	Activity last follow-up[b]	Assessment pain relief
1	F	arachnoiditis	9	90	43	7/26/89	sufentanil	catheter revision 8 mos	catheter revision 17 mos	catheter revision 30 mos	3.3	9	60	none
2	F	peripheral neuropathy	7	60	67	8/18/89	morphine				3.5	6	70	none
3	M	arachnoiditis	10	60	50	8/23/89	morphine				3.3	6	70	good
4	F	intercostal neuralgia	9	60	38	12/15/89	sufentanil	catheter revision 4 mos	pump replaced 22 mos	catheter revision 33 mos	3.0	5	89	good
5	F	arachnoiditis	7	69	55	1/16/89	morphine				3.9	4	80	good
6	M	peripheral neuropathy	7	70	77	6/11/90	morphine				1.5	7	70	none
7	F	arachnoiditis	9.5	70	73	7/31/190	morphine	pump-replaced 21 mos			2.5	5.5	60	good
8	F	phantom pain	8	60	62	12/14/90	morphine				0.5	6	60	fair
9	M	arachnoiditis	8	80	46	12/18/90	sufentanil	pump revision 9 mos			2	5	80	fair
10	F	intercostal neuralgia	10	80	44	1/7/91	sufentanil				0.7	7.5	80	fair
11	F	peripheral neuropathy	9	70	66	1/11/91	sufentanil				2.5	2	80	good
12	F	arachnoiditis	8	70	64	3/8/91	sufentanil				1.7	8	60	none
13	F	arachnoiditis	9	90	28	3/12/91	sufentanil	catheter revision 5 mos			1.0	9	90	none
14	F	arachnoiditis	8	70	25	8/29/91	sufentanil	catheter revision 1 mon			0.8	8	70	none
15	F	lumbosacral plexopathy	8	100	36	9/6/91	morphine				1.2	4	100	good
16	M	spinal cord injury	9	90	33	9/10/91	morphine				1.2	6	80	fair

[a] *VDS* verbal digital pain scale. [b] Activity rating: activity rating: 100 = Job, full-time, 90 = job, part-time, 80 = drives a car $\geq$ 1x/month, 70 = out of house and property $\geq$ 2x/month, 60 = out of house and property < 2x/month, 50 = out of house but not-off property, 40 = does household chores, 30 = no household chores but out of bed $\geq$ 6 hrs/day, 20 = out of bed < 6 hrs/day, 10 = bedbound, 0 = dead. [c] Length of time from implant to last follow-up or classification as longterm failure for pain relief.

patients had satisfactory pain relief with relatively low doses of opioid (morphine ≤ 1.0 mg/h, sufentanil ≤ 1.0 µg/h). A second group, however, required higher dose levels (morphine 1.5–2.0 mg/h, sufentanil 1.5–2.0 µg/h), but obtained satisfactory pain relief with long-term stable dosing. It was not possible to determine any predictive characteristics of the failure patients receiving an implanted pump.

For patients failing to obtain long-term pain relief, however, there was a steady dose escalation with decreasing analgesic benefits over 6–12 months after placement of the infusion pump. Drug holidays were tried in all of these patients. The patients were sensitive to lower doses of opioid after a drug holiday, however, in this series, there was a rapid dose escalation after a drug holiday and this resulted in a lack of adequate pain relief within 3–4 months.

Systemic opioid use continued in two-thirds of these patients. For half of these patients it consisted of occasional use of Schedule III narcotics. For the other half, however it consisted of the regular use (2–4 tablets per day) of Schedule II narcotics, usually oxycodone.

Three patients who failed to obtain long-term pain relief with either morphine or sufentanil, received intrathecal infusions of clonidine. These patients were started at a dose of 0.3 µg/h and increased to a maximum 50 µg/h. The pumps were programmed to increase in 0.1 µg/h increments reaching a maximal dose rate over a 12-week period. Two clonidine concentrations were used: 150 and 500 µg/ml. None of these three patients had any side effects including hypotension or hypertension. Two of the three patients experienced 25% pain reduction where intrathecal infusions of opioids had provided no pain relief. The patients have continued on the clonidine infusion with stable pain relief. The third patient experienced no pain relief despite a maximal dose of 50 µg/h.

Discussion

Spinal Cord Stimulation

Electrical stimulation has been applied to the spinal cord with either implanted or externalized systems since approximately 1967 [17]. The more recent availability of multichannel systems with electrode surfaces containing four contact points has significantly improved the ability to vary patterns of perceived stimulation [11].

Twenty-five years of experience with spinal cord stimulation has indicated a long-term success rates ranging from 50–70% [9–10]. Spinal cord stimulation has usually been found to be most effective for patients with neuropathic pain, especially unilateral with a radicular pattern in one leg. Until recently, treatment of midline back pain or bilateral leg pain has been considered more difficult with spinal cord stimulation because of limitations in the stimulator electrodes. The reason for this difficulty has been the small target areas in the epidural space of patients with such pain and the difficulties in adjusting the area of stimulation if there was any electrode movement. The patients in this series were thus stratified according to the pain.

Long-Term Intraspinal Infusion of Opioids

The intraspinal infusion of opioid on a long-term basis has been utilized on a frequent basis since approximately 1982 [4, 14]. It was initially applied to patients with cancer pain and thus limited survival times. In the last 5–8 years, this modality has been applied to patients with non-cancer pain [3, 6, 8]. The infusions have been performed either because of a failure of spinal cord stimulation during a testing phase or after permanent implantation of a stimulator system, or because of a hope that the opioid infusion might provide better long-term efficacy than long-term spinal cord stimulation.

Patients with cancer pain and expected survival times in excess of three to four months or with non-cancer pain have usually received implanted infusion pumps connected to an intrathecal catheter. The rationale has been to prevent infection associated with the long-term use of an externalized catheter as well as cost benefit considerations [1].

Reports of the long-term efficacy of spinal infusions using opioid agents have indicated moderate to marked pain relief in over 80% of cancer patients although dose escalations over survival times from 6–10 months have been noted [5, 12]. Limited reports of the use of this modality in non-cancer patients would suggest 25–50% minimum pain relief on a long-term basis in approximately 50–70% of patients with continued dose escalations in many of these patients over at a 24-month period [13]. There is an absence of data directly comparing spinal cord stimulation and long-term intraspinal infusion of opioids for patients with primarily neuropathic pain although it is generally agreed that neuropathic pain is less sensitive to opioids than nociceptive pain [15].

*Spinal Cord Stimulation and Spinal Infusion
of Opioid Comparisons*

The present review was undertaken to provide more information concerning the benefits as well as the disadvantages of these two modalities especially in relation to each other. It should be emphasized that this is a retrospective study. Although all patients had predominantly neuropathic pain, a stimulator system was offered preferentially to certain patients and a spinal infusion system to other patients based upon the pain pattern.

In the absence of other studies, however, the present review does provide information that can be useful in beginning to assess the relative roles of these two modalities. With the initial non-cancer application of the spinal infusion of opioids, it was hoped that this modality would become the preferred treatment for many patients. More recently, it has been thought that it might be a treatment option for patients who have failed to obtain relief with spinal cord stimulation.

The stratification of patients in this review to either spinal cord stimulation or long-term spinal infusion of opioid was based upon the pattern of pain because of limitations in existing stimulator hardware. Since this series, diamond-shaped electrode arrays and dual quadripolar systems have become available and could have been quite effective in many of these patients who were stratified to the spinal infusion of opioids.

Spinal cord stimulation in the patients in this review was fairly effective with $\geq 50\%$ pain relief in 62% of the patients. There was a significant subset of patients who did undergo a laminectomy for placement of a plate-type electrode. Although this is a more extensive operation, the long-term stability of these plate-type electrodes appears to be better. Experience from this series, however, did indicate that pain patterns in the chest area were not well treated with spinal cord stimulation because of discomfort caused by twitching of intercostal muscels. It was also difficult to obtain an adequate pattern of stimulation for patients with a minor component of low back and/or bilateral buttock pain.

For the long-term intraspinal infusion of opioids, the overall success rate was clearly less than that seen with spinal cord stimulation. Ony 38% of the patients experienced $\geq 50\%$ pain reduction and 25% experienced 25%–49% reduction. Dose escalation was noted in many of these patients although coverage of the pain patterns was very good and, by the nature of an infusion, patients were not affected by small catheter movements.

Comparison of the two groups clearly indicated that achieving and maintaining an adequate pattern of stimulation is the major concern with spinal cord stimulation while dose escalation is the major concern for spinal infusions. This experience would seem to indicate that seem to indicate that the spinal infusion of opioid on a long-term basis in non-cancer patients with neuropathic pain is best used as a last-resort option. It should be noted, however, that there was a patient subset in whom the long-term spinal infusions were very effective with stable low doses over a long peroid of time.

The use of clonidine in three of these patients who had failed to obtain long-term pain relief with morphine or sufentanil indicated a promising alternative agent that is being increasingly used in both Europe and Australia [2, 16, 18]. In the United States, it is available for spinal infusion under unvestigational approval from the Food and Drug Administration although it may be commercially available in the next 1–2 years. Experience from investigatous, including the present authors, indicates that it may be most effective for neuropathic pain sympathetic origin as in complex regional pain syndrome (CRPS) or reflex sympathetic dystrophy (RSD). Admixtures of morphine and clonidine may be more effective for sympathetically-independent, neuropathic pain. Future, possibly more effective infusion agents would include neuronal-specific calcium channel blockers or NMDA antagonists.

References

1. Bedder MD, Burchiel K, Larson A (1991) Cost analysis of two implantable narcotic delivery systems. J Pain Symptom Manag 6: 368–373
2. Du Pen S, Eisenach JC, Allin D, Zaccaro D (1993) Epidural clonidine for intractable cancer pain. Reg Anesth 18 [Suppl]: 23
3. Glynn C, Dawson D, Sanders R (1988) A double-blind comparison between epidural morphine and epidural clonidine in patients with chronic non-cancer pain. Pain 34: 123–128
4. Harbaugh RE, Coombs DW, Saunders RL, Gaylor M, Pageau M (1982) Implanted continuous epidural morphine infusion system. Preliminary report. J Neurosurg 56: 803–6
5. Hassenbusch SJ, Pillay PK, Magdinec M, Currie K, Bay JW, Covington EC, Tomaszewski MZ (1990) Constant infusion of morphine for intractable cancer pain using an implanted pump. J Neurosurg 73: 405–409
6. Hassenbusch SJ, Stanton-Hicks M, Soukup J, Covington EC, Boland MB (1991) Sufentanil citrate and morphine-in-bupivacaine as alternative agents in chronic epidural infusions for intractable non-cancer pain. Neurosurgery 29: 76–82
7. Littman GS, Walker BR, Schneider BE (1985) Reassessment of verbal and visual analog ratings in analgesic studies. Clin Pharmacol Ther 38: 16–23
8. Murphy TM, Hinds S, Cherry D (1987) Intraspinal narcotics: non-malignant pain. Acta Anaesthiol Scand 85 [Suppl]: 75–76

9. North RB (1993) The role of spinal cord stimularion in contemporary pain management. APS Journal 2: 91–99

10. North RB (1990) Spinal cord stimulation for intractable pain: long-term follow-up. J Spinal Disord 3: 356–361

11. North RB, Kidd DH, Zahurak M, James CS, Long DM (1993) Spinal cord stimulation for chronic, intractable pain: experience over two decades. Neurosurgery 32: 384–395

12. Onofrio B, Yaksh TL (1990) Long-term pain relief produced by intrathecal morphine infusion in 55 patients. J Neurosurg 72: 200–209

13. Paice JA, Penn RD, Shott S (1995) Intraspinal morphine for chronic pain: a retrospective study. J Pain Symptom Management

14. Poletti CE, Cohen AM, Todd DP, Ojemann RG, Sweet WH, Zervas NT (1981) Cancer pain relieved by long-term epidural morphine with permanent indwelling systems for self- administration. J Neurosurg 55: 581–4

15. Portenoy RK, Foley KM, Inturrisi CE (1990) The nature of opioid responsiveness and its implications for neuropathic pain: new hypotheses derived from studies opioid infusions. Pain 43: 273–286

16. Rauck RL, Eisenach JC, Jackson K, Young LD, Southern J (1993) Epidural clonidine treatment for refractory reflex sympathetic dystrophy. Anesthesiology 79: 1163–1169

17. Shealy CN, Mortimer JT, Reswick JB (1967) Electrical inhibition of pain by stimulation of the dorsal columns: preliminary clinical report. Anesth Analg 46: 489–491

18. Siddall PJ, Gray M, Rutkowski S, Cousins MJ (1994) Intrathecal morphine and clonidine in the management of spinal cord injury pain–a case report. Pain 59: 147–148

19. Zimmerman M (1984) Neurobiologic concepts of pain, its assessment and therapy. In: Bromm B (ed) Pain measurement in man. Neurophysiological correlates of pain. Elsevier, New York, pp 15–35

Correspondence: S. J. Hassenbusch, M.D., Ph.D., Department of Neurosurgery, C9.075, M.D. Anderson Cancer Center, 1515 Holcombe Boulevard, Houston, Texas 77030, U.S.A.

Acta Neurochir (1995) [Suppl] 64: 116–118

Treatment of the Failed Back Surgery Syndrome Due to Lumbo-Sacral Epidural Fibrosis

D. Fiume, S. Sherkat, G.M. Callovini, G. Parziale, and **G. Gazzeri**

Divisione di Neurochirurgia, Ospedale S. Filippo, Rome, Italy

Summary

The failed back surgery syndrome (FBSS) is a severe, long-lasting, disabling and relatively frequent (5–10%) complication of lumbo-sacral spine surgery. Wrong level of surgery, inadequate surgical techniques, vertebral instability, recurrent disc herniation, and lumbo-sacral fibrosis are the most frequent causes of FBSS. The results after repeated surgery on recurrent disc herniations are comparable to those after the first intervention, whereas repeated surgery for fibrosis gives only 30–35% success rate, and 15–20% of the patients report worsening of the symptoms. Computerized tomography (CT) with contrast medium and, in particular, Gd-DPTA enhanced MRI have recently allowed a differentiation between these two pathologies permitting us to adopt different therapies. In 1982–92 we applied spinal cord stimulation (SCS) as a first therapy of FBSS with proven lumbo-sacral fibrosis.

Fifty-five patients underwent percutaneous trial SCS with a mono/multipolar electrode placed at the level of Th9–12. In the 36 patients who had a positive response to the trial stimulation, the electrode was connected to an implantable neurostimulator. On January '94 a third party, not involved in the treatment of the patients, controlled 34 of the 36 patients with a mean follow-up of 55 months. We classified the patients reporting at least 50% pain relief and satisfaction with result as successful, and 56% of the patients fell in that category. 10 out of 34 patients were able to resume their work. The success rate was significantly higher in females (73%) than in males, and in radicular rather than axial pain. Our data have led us to consider SCS as a first choice treatment in FBSS due to lumbo-sacral fibrosis.

Keywords: Spinal cord stimulation; failed back surgery; pain.

Introduction

It is known that each year about one million patients all over the world undergo lumbo-sacral surgery for disc herniation. In spite of the improvement of surgical technique and particularly the adoption of micro-surgical approach, the success rate of these operations does not surpass 90–92% [3,4]. Failed back surgery is a syndrome characterized by severe, chronic and disabling pain which generally is resistant to physiotherapy and pharmacological treatment. Frequent causes for this condition are vertebral instability, surgery performed on wrong level, improper selection of patients, inadequate surgical technique, recurrent disc herniation and surgical complications. It is believed that the formation of epidural scar tissue is one of the most common causes.

Since the outcome of repeated surgery for epidural fibrosis generally is unsatisfactory [2, 6, 8, 9, 11, 15, 23, 25] there is a need for alternative treatment modalities. Since the late 70's spinal cord stimulation (SCS) has been regarded as being particularly effective for this condition [6, 10, 12, 20, 22].

Material and Methods

Our series includes 55 patients who between 1982 and 1992 were subjected to spinal cord stimulation for severe, disabling FBSS, unresponsive to conventional therapy. The average age was 52 years (36–66) and the length of the history of disease ranged from 1–12 years. Many of the patients had been submitted to repeated lumbosacral surgery (1.8 op/pat); 57% of the patients were males. Epidural fibrosis had been diagnosed with CT with contrast medium and, recently, with NMR with Gadolinium contrast.

Patients who had epidural fibrosis associated with a surgically treatable pathology such as canal or recess stenosis, vertebral instability or recurrent disc herniation were excluded. Also patients presenting with obvious signs of psychological disturbances were excluded. Patients presenting with radicular pain were preferred and those who had FBSS associated with pain only in the lumbar region were excluded. It is well known that it is difficult to produce paraesthesiae in the lumbar region and this can be the reason why the success rate for patients with pain confined to this area is low [5].

In all patients but one we used electrodes designed for percutaneous implantation (since 1987 we preferably used multipolar electrodes). The tip of the electrode was positioned at the level of Th9–10. Patients who reported more than 75% pain relief at the end of the trial stimulation period were permanently implanted with a multiprogrammable neurostimulator (Itrel II, Medtronic, Minn, USA). The stimulators were programmed with 85 Hz, 210 msec. We usually used stimulation with a cycle mode with 1 hour on–1 hour off.

The patients were evaluated in January '94 by a colleague who had not been involved in the treatment of the patients. Thirty-four of the 36 patients were interviewed and the patients filled in a questionnaire. One patient had died from myocardial infarction and one could not be retrieved. The mean follow-up was 55 months (1–10 years). Patients who reported at least 50% pain relief were considered as having a satisfactory outcome [8, 39]. The assessment was also based on the patient's own satisfaction with the treatment as well as on drug consumption and working ability.

Results

At the latest follow-up 56% of the patients were classified as being successful with more than 50% pain relief and with a self-reported satisfaction with the treatment. In 8% of the patients there was an equivocal success, since although these patients reported more than 50% pain relief they were not sure whether they would undergo again the SCS treatment knowing the outcome.

As has previously been reported by others [14, 20] we found a tendency towards a decreasing effect with time. However, it appeared that after an initial and moderate decrease of the effect of stimulation, the remaining amount of pain relief was retained throughout the subsequent years of follow-up.

In six patients (17%) the stimulators were removed due to infection in three cases and lack of satisfactory effect in another three. 73% of the patients continued to use the stimulator at the last control independently of their reported degree of pain relief. At the last control, 10 patients (29%) had stopped taking any drugs and 12 (35%) only used light analgesics. The changes of consumption of analgesic drugs are summarized in Table 1.

Twelve out of 28 patients who had been unable to work resumed their working activities after the treatment, 7 of them full time and 5 part time. Two patients who previously had been employed had to leave their jobs due to worsening of their symptoms. In total, 10 patients (29%) were believed to have been able to retain their work as a result of the treatment.

Table 1

Class	Drug consumption	Patients on admittance	Last follow-up
0	no	0	10
1	minor analgesics	3	12
2	major analgesics	24	7
3	morphine and derivates	7	5

There was no relationship between the success rate and the number of previous surgical interventions, duration and severity of the symptoms and time since the first operation. The success rate was significantly higher in females (69%) than in males (43%). It also appeared that patients with a predominance of radicular pain did better.

Complications

In six patients the electrodes were dislocated and in another three the electrodes had fractured. With the recent, improved electrodes such complications have been extremely rare. Five patients developed an infection which necessitated the removal of the system; two of these patients were reimplanted. In eight patients the neurostimulators had to be exchanged when the batteries were exhausted. In total, thirty-four patients needed 22 surgical revisions (64%). In no case was there any neurological complications.

Discussion

While the results of surgery for recurrent disc herniation are approximately the same as those of the first intervention [15, 25], the same can certainly not be said about repeated surgery for epidural fibrosis which generally has a poor outcome [6, 11, 15, 25]. According to Waddel [25] scar tissue removal only produces a new tissue injury which gives rise to deformation of new scar tissue. Johnston [9] ironically states that surgery for fibrosis may be a good exercise for the surgeon but it brings no benefit to the patient. Thomalske [23] reports complete recovery only in 36% of the patients operated on for epidural fibrosis in a series of 2000 operations for disc herniations. In order to provide effective treatment for complaints due to epidural fibrosis a reliable clinical and radiological differential diagnosis between recurrent disc herniation and epidural fibrosis is mandatory.

A free interval of more than one year after the last surgery has been suggested as indicating a recurrent disc herniation [8, 18]. Recently, a number of studies have emphasized the possibility of differentiating epidural fibrosis and recurrent disc herniation with the use of MRI enhanced with GdDTPA [2, 13, 19, 21]. In a series reported by Sotiropoulos [21] it was possible to identify with 100% certainty a disc herniation with MRI, later confirmed at surgical exploration. However, this author considered the diagnosis of fibrosis to be uncertain. On the contrary, Carella [2] claims that epidural fibrous tissue, without a mass effect, can be identified with T1 intermediate signal and hyperintense T2 using Gadolinium enhancement.

In the literature, inconsistent results of SCS treatment for pain due to epidural fibrosis have been presented: In some the outcome has found to be poor [14, 24], in some other excellent [5, 16, 17, 26]. No doubt, the employment of percutaneous trial stimulation, prior to the definite implantation, has markedly improved the long-term outcome [7].

Improved technology with multipolar leads and fully implantable neurostimulators has largely contributed to the reliability of SCS with less problems of equipment failure and electrode dislodgements. The 56% success rate in our series is comparable to the data recently reported by North [16] and De La Porte [5], and confirms the idea that SCS is preferable to repeated surgery in patients with pain due to epidural fibrosis. Apart from the effect on pain it was obvious that the treatment also resulted in a considerable reduction of consumption of analgesics in that 64% of the patients at follow-up used no drugs or only peripherally acting, light analgesics. It was also obvious that the working capacity was improved in that 29% of the patients were able to go back to their previous work. However, the great number of surgical interventions for revision of the system (22 procedures in 34 patients) was a problem.

In our experience, provided a correct diagnosis of FBSS can be established, SCS offers a better chance of helping the patients than repeated surgery. Moreover, this treatment modality is not burdened by any severe complications or side-effects, although repeated surgery is sometimes necessary and the equipment is still costly.

References

1. Burton CV (1978) Lumbosacral arachnoiditis. Spine 3: 24–30
2. Carella A, Andreula CF (1990) Impiego del Gd-DTPA nello studio con risonanza magnetica nella patologia degenerativa del rachide lombare. Riv Neurobiol 3 [Suppl 3]: 91–98
3. Caspar W, Campbell B, Barbier D *et al* (1990) The Caspar microsurgical discectomy and comparison with a conventional standard lumbar disc procedure. Neurosurgery 28: 78–87
4. Davis RA (1994) A Long-term outcome analysis of 984 surgically treated herniated lumbar discs. J Neurosurg 80: 415–422
5. De La Porte C, Siegfried J (1993) Spinal cord stimulation in failed back surgery. Pain 52: 55–61
6. De La Porte C, Siegfried J (1983) Lumbosacral spinal fibrosis (spinal arachnoiditis): its diagnosis and treatment by spinal cord stimulation. Spine 8: 593–603
7. Erickson DL (1975) Percutaneous trial of stimulation for patient selection for implantable stimulating devices. J Neurosurg 43: 440–444
8. Finnegan WJ, Fenlin JM, Maevel JP *et al* (1979) Results of surgical intervention in the symptomatic multiply-operated back patient. J Bone Joint Surg 61–A: 1077–1082
9. Johnston J, Matheny J (1978) Microscopic lysis of lumbar adhesive arachnoiditis. Spine 3: 36–39
10. Kumar K, Nath R, Wyant GM (1991) Treatment of chronic pain by chronic pain by epidural spinal cord stimulation: a 10 year experience. J Neurosurg 75: 402–407
11. Law JD, Lehman RAW, Kirsch WM (1978) Reoperation after lumbar intervertebral disc surgery. J Neurosurg 48: 259–263
12. Long DM, Erickson DE (1975) Stimulation of the posterior columns of the spinal cord for relief of intractable pain. Surg Neurol 4: 134–141
13. Masaryk TJ (1989) Herniated nucleus pulposus vs epidural fibrosis: disk vs scar. MRI Decisions 30: 20–26
14. Meglio M, Cioni B, Rossi GF (1989) Spinal cord stimulation in management of chronic pain: a 9-year experience. J Neurosurg 70: 519–524
15. North RB, Campbell JN *et al* (1991) Failed back surgery syndrome: 5 year follow-up in 102 patients undergoing repeated operation. Neurosurgery 28: 685–691
16. North RB, Ewend MG, Lawton MT *et al* (1991) Failed back surgery syndrome: 5 years follow-up after spinal cord stimulator implantation. Neurosurgery 28, 1991: 692–699
17. North RB, Kidd DH, Zahruk M *et al* (1993) Spinal cord stimulation for chronic, intractable pain: experience over two decades. Neurosurgery 32: 384–395
18. Rothman RH, Bernini PM (1981) Algorithm for salvage surgery of the lumbar spine. Clin Orthop 154: 14–17
19. Schubiger O, Valvanis A (1982) CT differentiation between recurrent disc herniation and postoperative scar formation: the value of contrast enhancement. Neuroradiology 22: 251–254
20. Siegfried J, Lazorthes Y (1982) Long-term follow-up of dorsal column stimulation for chronic pain syndrome after multiple lumbar operation. Appl Neurophysiol 45: 201–204
21. Sotiropoulos S, Ghafetz NI, Lang P *et al* (1989) Differentiation between postoperative scar and recurrent disk herniation: prospective comparison af MR, and contrast-enhanced CT. AJNR 10: 639–643
22. Sweet W, Wepsic J (1974) Stimulation of the posterior columns of the spinal cord pain control. Clin Neurosurg 21: 278–310
23. Thomalske G, Galow W, Ploke G (1977) Critical comments on a comparison of two series (1000 patients each) of lumbar disc surgery. Adv Neurosurg 4: 22–27
24. Urban BJ, Nashold Jr BS (1978) Percutaneous epidural stimulation of the spinal cord for relief of pain. J Neurosurg 48: 323–328
25. Waddel G, Kummel EG, Lotto WN, *et al* (1979) Failed lumbar disc surgery and repeat surgery following industrial injuries. J Bone Joint Surg (A) 61: 201–207
26. Winkelmuller W (1981) Experience with the control of low back pain by the dorsal column stimulation (DCS) system and by the peridural electrode system (Pisces). In: Hosobuchi Y, Corbin T (eds) Indications for spinal cord stimulation. Excerpta Medica, Amsterdam, pp 34–41

Correspondence: Dario Fiume, M.D., Via Casale Ghella 45, 1-00189 Rome, Italy.

Acta Neurochir (1995) [Suppl] 64: 119–124
© Springer-Verlag 1995

Significance of the Spinal Cord Position in Spinal Cord Stimulation

J. Holsheimer[1], **G. Barolat**[2], **J.J. Struijk**[1], and **J. He**[2]

[1]Institute for Biomedical Technology, University of Twente, Enschede, The Netherlands and [2]Department of Neurological Surgery, Jefferson Medical College, Philadelphia, PA, U.S.A.

Summary

The effects of the antero-posterior and medio-lateral positions of the spinal cord in the dural sac on the perception threshold and paresthesia coverage in spinal cord stimulation were analyzed.

The distributions of the dorsal cerebrospinal fluid (CSF) layer thickness, measured from transverse MR scans of normal subjects at various spinal levels, were used to calculate the distributions of threshold voltages for the stimulation of spinal nerve fibers by a computer model. These theoretical threshold distributions were shown to fit well to the corresponding distributions of perception threshold measured in patients.

It is concluded that the thickness of the dorsal csf layer is the main factor determining the perception threshold and paresthesia coverage in spinal cord stimulation: an increasing thickness raises the threshold and reduces the coverage, and vice versa.

The effects of an asymmetrical electrode position with respect to the spinal cord midline were also analyzed by computer modeling. It is concluded that a lateral asymmetry of less than 1 mm gives a significant reduction of perception threshold and may result in unilateral paresthesiae.

Keywords: Spinal cord stimulation; perception threshold; paresthesia coverage; spinal cord position; computer modeling.

Introduction

In spinal cord stimulation (SCS) the perception threshold (Tp) is defined as the lowest stimulus level (in Volts) at which a patient feels paresthesia. It is well known that Tp varies strongly, both as a function of the vertebral level of the electrode and among patients having the electrode at the same level [1, 4].

Physically it is expected that Tp increases as the distance between the (epidural) electrode and the spinal cord is increased. This distance D is almost identical to the thickness of the dorsal cerebrospinal fluid (CSF) layer. The relationship of D and Tp was shown by calculating the applied electrical field in the spinal cord and its effect on spinal nerve fibers by computer models [3, 5, 7, 9, 11].

It was also shown by these models that the threshold voltage to stimulate large dorsal root (DR) fibers is generally less than the threshold voltage of the corresponding dorsal column (DC) fibers [3,11]. Moreover, the model predicts that the DC fiber threshold increases more steeply than the DR fiber threshold as D is increased [7]. Therefore, we assume that Tp will generally be related to the activation of large DR fibers.

An important clinical aspect is the paresthesia coverage related to the stimulation of DR fibers and DC fibers. A segmentary paresthesia will be obtained if only DR fibers in a few dorsal rootlets can be activated within the usage range, whereas a broader paresthesia coverage is possible if DC fibers are stimulated as well.

Although the distance D seems to be an important parameter for Tp, other factors may also contribute to the large variation of D. In order to investigate the contribution of D we analyzed the correlation of D and Tp. The best way to do so is to use measurements of both Tp and D of the same group of patients. However, such clinical data are not available. Therefore, we used Tp data from the Division of Functional Neurosurgery of Thomas Jefferson University, Philadelphia [4] and D values obtained from MR scans of normal subjects at the Medical Spectrum Twente Hospital, Enschede [6]. By calculating DR fiber threshold voltages (Tcal) for all values of D we were able to compare the means of Tcal and Tp and their distributions for stimulation at several vertebral levels.

The effect on Tp of an electrode position lateral to the spinal cord midline was analyzed on basis of both clinical data and computer modeling. The effects of D, Tp and asymmetrical stimulation on paresthesia coverage are discussed.

Methods

Measurement of Perception Threshold (Tp)

Tp values were determined in 136 chronic pain patients and sorted according to the vertebral level of the cathodal contact of the SCS electrode (Resume, Medtronic Inc, Minneapolis, MN). During the measurement of Tp at a constant pulsewidth (210 μs) and rate (50 pps) the patients were in a supine position [4].

The patients were affected from chronic pain due to either the "failed back surgery" syndrome or neuropathic pain. None of the patients were found to have any anatomical abnormality near the SCS electrode, neither a central or peripheral neuropathology that could affect their ability to perceive paresthesia.

Measurement of Dorsal CSF Layer Thickness (D)

Transverse MR scans were made at midcervical (C4–C6), midthoracic (T5–T6) and lowthoracic (T11–T12) vertebral levels of 26 normal subjects in a supine position. Strongly T2 weighed Turbo Spin Echo scans were used to obtain good contrast between spinal cord, csf and dura mater. Transverse sections were 5.0 mm thick and were separated by 5.5 mm.

Transverse MR images at T5 and C5 are shown in Figs. 1a and 1b, respectively. The thickness of the dorsomedial csf layer (D) was measured from the enlarged images and for each subject the values from 2 or 3 adjacent sections at each vertebral level were averaged. For details see Holsheimer et al. [6].

Calculation of Threshold Stimuli of Dorsal Root Fibers (Tcal)

Computer models consisting of two parts were used: a volume conductor model and a myelinated nerve fiber model. The 3-dimensional volume conductor models represent the gross anatomy of a midcervical, a midthoracic and a lowthoracic spinal segment and are based on MR scans. A transverse section of the midcervical model is shown in Fig. 1c. The models include the white and grey matter of the spinal cord, csf, dura mater, epidural fat, vertebral bone and surrounding tissues and two neighboring contacts of the Resume electrode as well as the corresponding electrical tissue conductivities. The dorsal csf layers D of the midcervical, midthoracic and lowthoracic models were 2.4 mm, 5.8 mm and 3.6 mm, respectively. Usually the contacts were centered at the spinal cord midline. The transverse size of the models was approximately 25 × 25 mm and their lenght 60 mm.

The cathode and the anode were set at different voltages and the voltage distribution in the model was computed. Subsequently, the voltages corresponding to the positions of the Ranvier nodes of a 15 μm DR fiber entering the spinal cord near the cathode were applied to the cable model of a myelinated nerve fiber. Then the threshold stimulation voltage for its excitation (Tcal) was computed, using a 210 μs pulse. Tcal was considered to mimic Tp. For details see Struijk et al. [9,10,11].

Results

Thickness of the Dorsal CSF Layer

From 26 male subjects (19–38 years) the mean thickness D was smallest at C4–C6 (2.5 mm) and largest at T5–T6 (5.8 mm), whereas the standard deviations (sd) were 0.8 mm and 1.8 mm, respectively. At T11 the mean value was 3.6 mm (sd = 1.6 mm) [6]. At most levels D had a skew (non-Gaussian) distribution, as shown in Figs. 2a and 2b for C4–C6 and T5–T6, respectively.

Perception Threshold and Spinal Level of Stimulation

Tp values selected for this study were from bipolar combinations of neigboring contacts of Resume electrodes (6 mm separation) placed dorsomedially in the areas C4–C6, T4–T7 and T10–T11. The mean Tp was lowest at C4–C6 (0.67 V, sd = 0.33 V, n = 29) and highest at T4–T7 (1.70 V, sd = 0.93 V, n = 24). The mean value at T10–T11 was 1.16 V (sd = 0.60 V, n = 51).

The means and standard deviations of Tp in the three spinal areas are presented in Fig. 3 (right side error bars). It is shown that the mean Tp increases as the mean D is increased. The large standard deviations of Tp are due to the large variations of D (Fig. 2), as will be shown below.

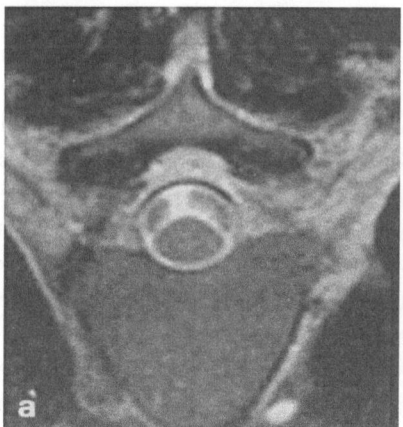
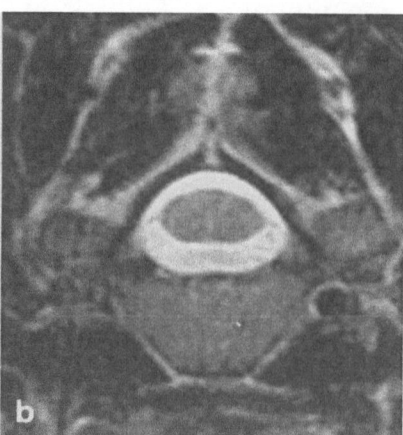
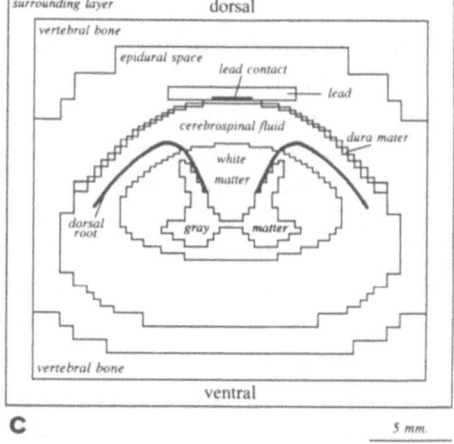

Fig. 1. Transverse Turbo Spin Echo MR scans at vertebral levels T5 (a) and C5 (b); transverse section of midcervical computer model (c); posterior side up

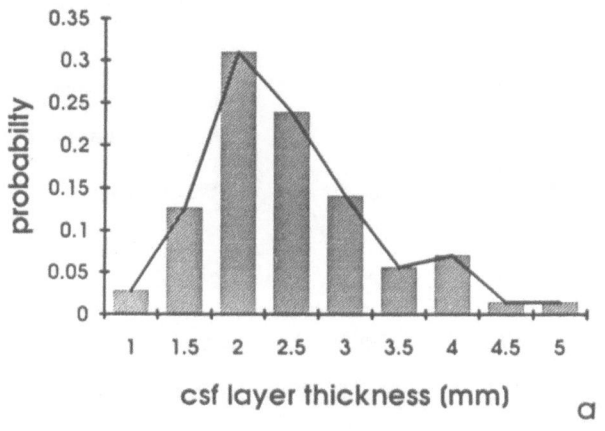

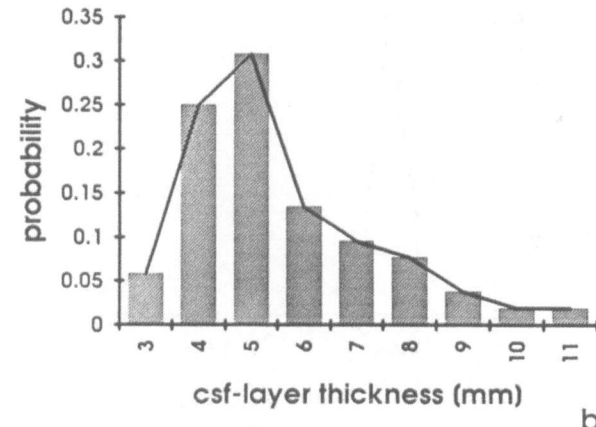

Fig. 2. Probability distributions of D at C4–C6 (a) and T5–T6 (b), measured from transverse MR scans

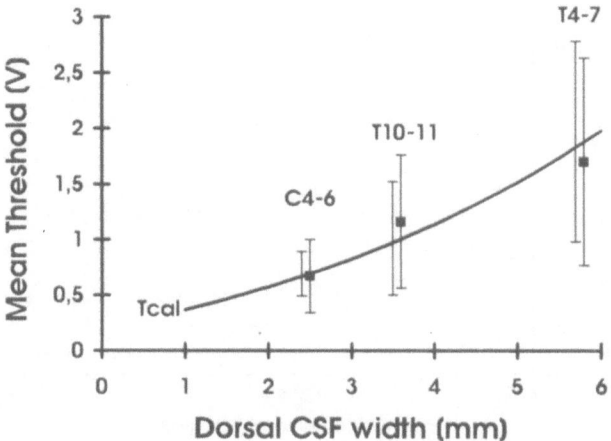

Fig. 3. Tcal: curve showing the theoretical relation of D and Tp, calculated from the computer model; left side error bars: standard deviations of Tcal at C4–C6 (D = 2.4 mm), T11 (3.6 mm) and T5–T6 (5.8 mm) calculated from the sd values of D; right side error bars: means and sd values of Tp at C4–C6, T10–T11 and T4–T7; bipolar stimulation with narrowly separated contacts (Resume)

Calculated Threshold Stimuli of Dorsal Root Fibers

Bipolar contacts of 3.6 × 3.6 mm, separated longitudinally by 6.6 mm and placed medially in the dorsal epidural space of the 3D models were used to compute Tcal for D values ranging from 1.6 mm to 7.2 mm. For any value of D the Tcal values were almost identical when the midcervical and the midthoracic models were used.

The relation of Tcal (V) and D (mm) fitted the following exponential function (corr. coefficient = 0.999)

$$Tcal = 1.99*10^{0.087D} - 1.48 \qquad (1)$$

In a previous paper we concluded that Tcal values were 2.5 – 3 times higher than the corresponding Tp

values [11]. Accordingly, we modified (1) proportionally (1/2.6) into

$$Tcal = 0.765*10^{0.087D} - 0.57 \qquad (2)$$

This theoretical relation is presented by the curve Tcal in Fig. 3. Moreover, the standard deviations of Tcal, calculated from the standard deviations of D at C4–C6, T5–T6 and T11, are shown at the left side of the error bars indicating the standard deviations of Tp. It is shown that both the mean Tp values and their standard deviations correspond well with the computed values, taking into account the small numbers of experimental data, ranging from 21 to 51.

However, at the various spinal levels D, Tp and Tcal do not have normal distributions and therefore, the relation of Tp and Tcal is not sufficiently determined by their means and standard deviations.

Distributions of Measured and Calculated Thresholds

The distributions of Tcal were obtained by transforming the distributions of D at C4–C6 and T5–T6 (Fig. 2) with the nonlinear relation of formula 2. The probability distributions of Tcal and Tp at C4–C6 are presented in Fig. 4a and those at T4–T7 in Fig. 4b. Like the probability distributions of D (Fig. 2), those of Tp and Tcal are skew. Figs. 4a and 4b show that Tp and Tcal have similar distributions.

A quantitative comparison of the corresponding distributions of Tp and Tcal was made by calculating the 25th and 75th percentiles (the median values, or 50th percentiles, were fitted). These percentiles, given in Table 1, affirm the similarity of the probability distributions of Tcal and Tp shown in Figs. 4a and 4b. The distributions of Tp are slightly wider than those of Tcal (see Discussion).

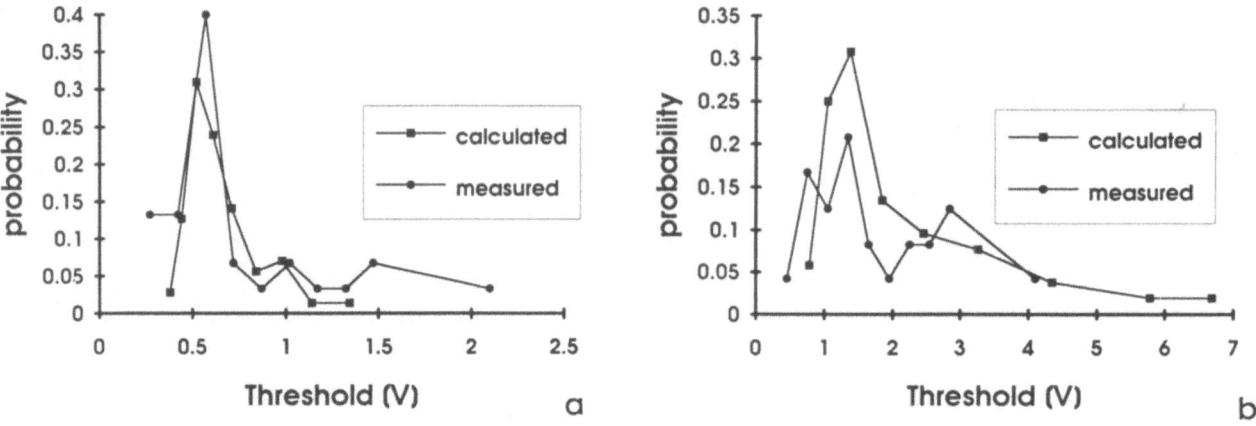

Fig. 4. Probability distributions of measured (Tp) and calculated (Tcal) thresholds at C4–C6 (a) and T4–T7 (b); bipolar stimulation with narrowly separated contacts (Resume)

Table 1. *Percentile Values of Threshold Distributions (volts)*

	Midcervical			Midthoracic		
	25th	50th	75th	25th	50th	75th
Tp	0.46	0.58	0.75	1.06	1.43	2.38
Tcal	0.50	0.58	0.70	1.18	1.43	2.15

Perception Threshold and Mediolateral Electrode Position

Recently we have shown that Tp is reduced significantly when the electrode is 2 mm or more off the radiological midline [4]. The mean reduction of Tp was 40% when the electrode was at C6–T1, whereas at T8–T10 and T11–T12 the mean reductions were 25% and 31%, respectively.

For various mediolateral positions of the Resume electrode with respect to the spinal cord midline stimulation thresholds of DR fibers were calculated using the computer models. In Fig. 5 Tcal of the left DR fiber is given as a percentage of its value with the electrode at midline (= 100%). The electrode was moved up to 1.8 mm to the left and right sides in the midcervical (C5) and the midthoracic (T5) models. It is shown that Tcal is reduced increasingly as the electrode is moved to the left side, whereas it is increased as the electrode is moved in the opposite direction. The latter also represents the change of Tcal of a right side DR fiber when the electrode is moved to the left side. It is also shown in Fig. 5 that the changes of Tcal in the C5 model are larger than in the T5 model, due to the smaller dorsal csf layer (2.4 mm and 5.8 mm, respectively). Because the computer models predict that Tp is usually related to stimulation of DR fibers, Tp will be reduced as the

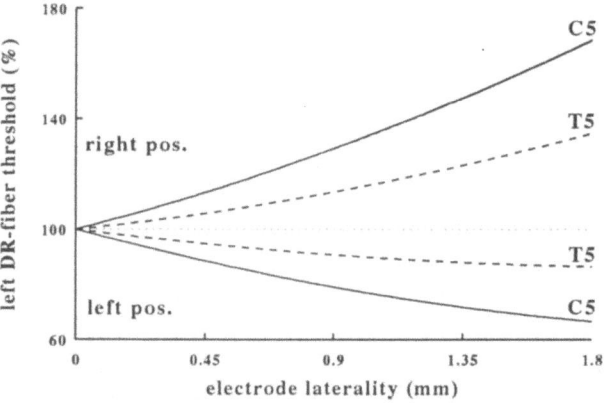

Fig. 5. Calculated changes (%) of left side DR fiber threshold as the electrode is moved from the spinal cord midline to 1.8 mm lateral (right and left sides) at levels C5 (D = 2.4 mm) and T5 (D = 5.8 mm); bipolar stimulation with narrowly separated contacts (Resume)

electrode is moved from a position corresponding to the spinal cord midline to a lateral position. When the center of the electrode is 1.8 mm lateral, Tp will be reduced by 32.9% when D = 2.4 mm (midcervically) and by 13.6% when D = 5.8 mm (midthoracically). Similarly, the threshold voltage to stimulate fibers in the opposite dorsal root will be increased by 68.2% and 34.5%, respectively.

In general this model prediction is in accordance with the conclusions from clinical data mentioned at the start of this section [4]. When stimulating at C6–T1, having the smallest mean D, the mean reduction of Tp is largest (40%). In contrast, at T8–T10 the mean D is largest and the mean reduction of Tp is smallest (25%). The clinical and modeling data could not be compared quantitatively, because the values of D related to the Pt values were not available and no selection of only narrow bipolar combinations of the Resume electrode was made.

Conclusions and Discussion

The relationship of D and Tp has been analyzed by comparing Tp values measured from a group of patients and the computed thresholds (Tcal) based on D values measured from another group of subjects. Tp data were obtained from bipolar stimulations with narrowly separated contacts of the Resume electrode (Medtronic Inc, Minneapolis, MN), whereas Tcal values were calculated by using the same stimulation in the computer models. Similar as in a previous study using non-selected Tp data [4] we have shown that the mean Tp at various spinal levels correlates well with the mean D, taking into account the large variations of D and Tp. Moreover, the probability distributions of D and Tp at various vertebral levels are correlated as well. Therefore, D can be considered the main factor determining Tp, probably by the exponential relation calculated from the computer models. Other factors increasing Tp are the presence of epidural fat or scar tissue between the electrode and the dura mater and an abnormally thick dura mater near the electrode.

Stimulation of only DR fibers in some dorsal rootlets will result in a small, segmentary paresthesia coverage. However, when the dorsal columns are stimulated a larger paresthesia coverage can be obtained. Because the computer model predicts that the threshold for stimulating DC fibers increases far more than the threshold for DR fibers when D is increased [7], the probability of stimulating DC fibers at large values of D is very small when stimulating with any contact combination of currently available SCS electrodes. Therefore, a large D (midthoracically) will result in a high Tp and small, usually segmentary paresthesias, as is well known from clinical practice. Conversely, a smaller D (midcervically, lowthoracically) results in a lower Tp and broader paresthesia coverage. This trend in paresthesia coverage was shown by Barolat et al. [2] who presented the quantified topographical distributions at almost maximum stimulation at many levels. The small numbers of data related to stimulation at C7–T5, however, were not included. Stimulation at these levels was usually avoided due to the small (segmentary) paresthesia coverage [1].

The relation of Tp and paresthesia coverage described only holds for a midline position of the electrode, resulting in symmetrical paresthesiae. An asymmetrical electrode position will result in a reduction of both Tp and paresthesia coverage, the latter being asymmetrical or even unilateral. Clinical data on the reduction of Tp were presented in a previous paper [4], whereas Barolat et al. [2] concluded that a reduction of paresthesia coverage is generally obtained when the electrode is 3 mm or more off the radiological midline.

The occurrence of unilateral paresthesiae is due to the simultaneous reduction of DR fiber thresholds at one side, the increase of DR fiber thresholds at the opposite side and the increase of DC fiber thresholds, as compared to the values when the electrode is at the spinal cord midline position. Taking into account the limited range of stimulation, having an average maximum of $\simeq 40\%$ beyond Tp [4, 8], the computer model predicts that paresthesia will be unilateral when the electrode is only slightly off the spinal cord midline. According to Fig. 5 a difference of 40% between the left and right DR fiber thresholds will be obtained when the electrode is $\simeq 0.7$ mm lateral with D = 2.4 mm and $\simeq 1.5$ mm with D = 5.8 mm. In order to obtain bilateral paresthesiae in midcervical SCS the mediolateral distance between the center of the electrode and the spinal cord midline should thus on the average be less than 0.7 mm and still smaller when paresthesiae should be symmetrical.

The MR study has shown that in about 40% of the subjects the spinal cord midline and the vertebral midline were 1–2 mm apart at all levels investigated [6]. This result and its calculated effect on paresthesia distribution is in accordance with Barolat et al. [1] who reported that the percentage of paresthesiae felt symmetrically when the stimulating contacts were perfectly located at the radiological midline was only 27%. These results clearly show that the radiological midline is an unreliable reference for SCS electrode placement. They also indicate that the difference between paresthesia coverages related to electrode positions more and less than 3 mm from the radiological midline [2] will be less pronounced, since a radiological laterality of e.g. 3 mm can be related to a laterality of 1–5 mm with respect to the spinal cord midline.

The variation of the mediolateral spinal cord position among subjects will cause an extra variation of Tp and a reduction of its mean value, although these effects will be relatively small in comparison to the influence of the variation of D. Indeed, we found that Tp had a slightly wider distribution than Tcal (Table 1).

Acknowledgement

This investigation was supported by a grant from Medtronic Inc., Minneapolis, MN.

References

1. Barolat G, Zeme S, Ketcik B (1991) Multifactorial analysis of epidural spinal cord stimulation. Stereotact Funct Neurosurg 56: 77–103
2. Barolat G, Massaro F, He J, Zeme S, Ketcik B (1993) Mapping of sensory responses to epidural stimulation of the intraspinal neural structures in man. J Neurosurg 78: 233–239
'3. Coburn B (1985) A theoretical study of epidural electrical stimulation of the spinal cord. Part II: effect on long myelinated fibers. IEEE Trans Biomed Eng 32: 978–986
4. He J, Barolat G, Holsheimer J, Struijk JJ (1994) Perception threshold and electrode position for spinal cord stimulation. Pain 59: 55–63
5. Holsheimer J, Struijk JJ (1991) How do geometric factors influence epidural spinal cord stimulation? A quantitative analysis by computer modeling. Stereotact Funct Neurosurg 56: 234–249
6. Holsheimer J, den Boer JA, Struijk JJ, Rozeboom AR (1994) MR assessment of the normal position of the spinal cord in the spinal canal. Am J Neuroradiol 15: 951–959
7. Holsheimer J, Struijk JJ, Tas NR (1995) Effects of electrode geometry and combination on nerve fibre selectivity in spinal cord stimulation. Med Biol Eng Comp: in press
8. Jobling DT, Tallis RC, Sedgwick EM, Illis LS (1980) Electronic aspects of spinal-cord stimulation in multiple sclerosis. Med Biol Eng Comp 18: 48–56
9. Struijk JJ, Holsheimer J, van der Heide GG, Boom HBK (1992) Recruitment of dorsal column fibers in spinal cord stimulation: influence of collateral branching. IEEE Trans Biomed Eng 39: 903–912
10. Struijk JJ, Holsheimer J, Boom HBK (1993a) Excitation of dorsal root fibers in spinal cord stimulation: a theoretical study. IEEE Trans Biomed Eng 40: 632–639
11. Struijk JJ, Holsheimer J, Barolat G, He J, Boom HBK (1993b) Paresthesia thresholds in spinal cord stimulation: a comparison of theoretical results with clinical data. IEEE Trans Rehab Eng 1: 101–108

Correspondence: J. Holsheimer, Ph.D., Department of Electrical Engineering, University of Twente, P.O. Box 217, 7500 AE Enschede, The Netherlands.

Acta Neurochir (1995) [Suppl] 64: 125–127

Anatomical Findings in Microsurgical Vascular Decompression for Trigeminal Neuralgia. Correlations Between Topography of Pain and Site of the Neuro-Vascular Conflict

M. Sindou, M. Chiha, and **P. Mertens**

Department of Neurosurgery, Neurological Hospital P. Wertheimer, University of Lyon, France

Summary

We here report on the anatomical findings in a series of 350 patients with trigeminal neuralgia (TN) and operated on using a microsurgical key-hole approach to the CPA.

In 5.7% there was a tumour or a vascular malformation, in 2.3% a mega-vertebro-basilar-artery. Among the remaining 322 (= real idiopathic TN), only 3.1% had no visible compressive factor, whilst 96.9% had one (or several) conflicting vessel(s): SCA in 90%, AICA in 23.6%, a vein in 24.7%. In 35.7% of the patients, several neuro-vascular conflicts (NVC) were found. Beside the NVC(s), a global atrophy of the entire root was seen in 67% of the cases. Degree of severity of the NVC and its site along the root were studied. The site of the conflict was: anteriorly to the root when pain was in V1, anteriorly and superiorly when in V2, superiorly and posteriorly when in V3.

Keywords: Micro-vascular decompression; neuro-vascular conflict; pain; trigeminal neuralgia.

Introduction

Thanks to the pioneering works of Dandy [1], Gardner [2], and Jannetta [3], the notion of a neuro-vascular conflict (NVC) as an important etiological factor in idiopathic trigeminal neuralgia (TN) has become widely accepted. Although a large number of series have been published since then, few reports with detailed operative findings are available.

Since 1979, 1300 patients referred for TN resistant to prolonged high doses of carbamazepine were operated on by the senior author (MS). 950 had a percutaneous thermo-rhizotomy and the other 350 a microsurgical key-hole retromastoid supracerebellar approach [4]. The anatomical-pathological findings observed under the operative microscope were carefully noted and drawn.

Material

This series of 350 patients consisted of 53% females and 47% males. Age ranged from 24 to 84 years, 60 on average. 60% had pain on the right side and 40% on the left. V1 division was affected in 32%, V2 in 76.5% and V3 in 62% (in a majority of the patients more than one division was involved). The mean duration of the pain history was 7 years. In 56% of the cases TN was of the paroxysmal type, i.e. typical; in 44% it was atypical in the sense that there was a permanent burning pain component associated with the paroxysmal crises.

Anatomical Findings

Etiology and Results of Surgery

In 20 cases (5.7%) a tumour or a vascular malformation was found. In 7 of them, TN was typical, without any deficit. In all 20, the lesion was successfully removed with total relief of pain.

In 8 cases (2.3%) there was an atherosclerotic mega-vertebro-basilar artery. These patients had an atypical TN with a permanent burning component. They underwent a selective juxta-pontine rhizotomy of pars major, due to the impossibility of pushing away the compressive mega-artery. Relief was total in all, but 3 cases suffered a moderate hypoesthesia dolorosa.

Of the group of 322 patients with a true idiopathic TN, 10 cases (3.1%) had no vascular conflict. These ten patients had no characteristic clinical features compared to the group with a conflicting vessel, but their nerves were globally atrophic. These ten patients were treated by a partial section of pars major. A complete relief was achieved in all except one patient. Numbness occurred in all, and one suffered a severe anesthesia dolorosa.

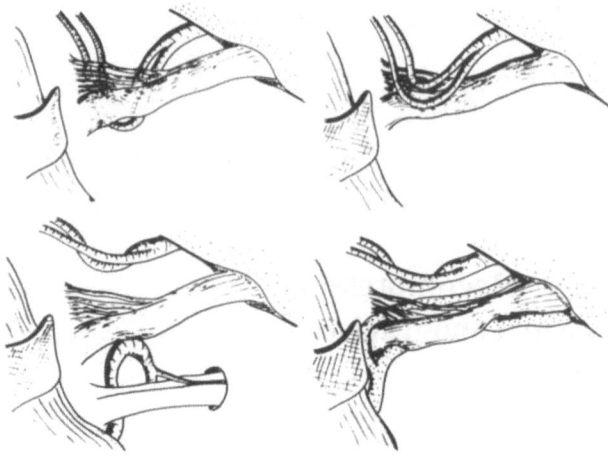

Fig. 1. Examples of various types of Neuro-vascular conflicts (seen through a right posterior CPA approach): SCA in anterior (= medial) (upper left) or superoposterior (= supero-lateral) (upper right) position; AICA inferiorly cross-compressing the Vth REZ(lower left). Satellite trigeminal veins embedded in nerve tissue (lower right)

In the remaining 312 patients (96.9% of the idiopathic group), one (or several) conflicting vessel(s) was identified (examples in Fig. 1): a superior cerebellar artery in 90%; an anterior inferior cerebellar artery in 23.6%; a vein embedded into the nerve in 24.7%; a basilar artery in 3.2%. Of prime importance was the fact that several conflicting vessels were found in 35.7% of the patients.

In twelve patients a partial section of the pars major was performed because it was not possible to decompress the root satisfactorily. Of these, 9 enjoyed good relief. Of the 300 patients treated with MVD, 2 (0.6%) died because of an hemorrhagic infarction of the cerebellum, 7 (2.3%) had an immediate failure (for unknown reasons) and they underwent thermorhizotomy, 250 (83.8%) had total relief, 25 (8.3%) suffered some episodic crises but did not require medication, 12 (4%) had incomplete improvement requiring additional medical treatments, and 4 (1.3%) had a "dissociated" cure (with complete relief of paroxysmal crises but not of the permanent burning component). Of prime importance is the fact that among the patients who had total relief, the effect was obtained immediately after surgery in 75% of the patients and after a few weeks or months in the remaining 25%.

Of the 291 patients with an effective MVD, 14% had a recurrence after a 1 to 12 years' follow-up (5 yrs on average). In three-fourth of them the recurrence occurred within the first post-operative year.

Characteristics of the Neuro-Vascular Conflicts

Degree of severity of the conflict. When several offending vessels were found, only the one causing the major conflict was considered for the study. The degree of severity was graded as 1) when the vessel was in contact with the root but without any visible indentation (18.6% of the cases), 2) when there was a displacement and/or distorsion of the root (46.4%), 3) when a significant indentation in the root was present (35%).

Site of the conflicts along the root. All the conflicts, if several in the same patient, were recorded. The following sites were found: at the trigeminal root entry zone in 78%, in the midthird of the root in 40%, and at the exit of the root from the Meckel's cave in 12%.

Location of the conflicts around the root (Fig. 2). Only the main vascular conflict, if several in the same patient, was considered for this study. The location was medial or medio-superior to the root in 59%, superior or supero-lateral in 33%, and latero-inferior or inferior in 8% of the patients.

Root Alterations and Surrounding Abnormalities

Beside the focal lesions due to the conflict, alterations of the whole trigeminal root were frequently observed. In 45% the root was significantly atrophic and in 22% there was a marked atrophy.

In addition to these findings, the following abnormalities were found: a significant degree of local arachnoiditis adhesive to the root in 22%, an absence of cisterns around the nerve due to small size of the posterior fossa in 5%, and a marked angulation of the root at its junction with the triangular plexus, when crossing over the petrous ridge after its exit from Meckel's cave, in 19%. Such an angulation frequently coexisted with a SCA in a superior position, strongly pushing down the nerve, making it hammock-shaped and markedly atrophic.

Correlations Between Topography of Pain and the Site of the Conflict Around the Root

Table 1 shows the correlations between the topography of pain with regard to the trigeminal division(s)

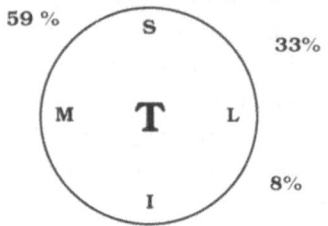

Fig. 2. Location of the conflict around the root

Table 1. *Correlation Between Topography of Pain and Site of the Vascular Conflict Around the Root*

Topography of trigeminal pain	Site of the vascular conflict around the root			
	Ant.	Sup.	Post.	Inf.
V1:4	4 (100%)			
V1–V2:38	29 (76.3%)	9 (23.7%)		
V1–V2–V3:17	10 (58.8%)	6 (35.2%)	1 (6%)	
V2:33	19 (57.5%)	11 (33%)	2 (6.1%)	1 (3.1%)
V2–V3:37	20 (54%)	14 (37.8%)	1 (2.8%)	2 (5.4%)
V3:19	6 (31.6%)	8 (42.2%)	3 (15.7%)	2 (10.5%)
Total cases: 148	88 (59.5%)	48 (32.4%)	7 (4.7%)	5 (3.4%)

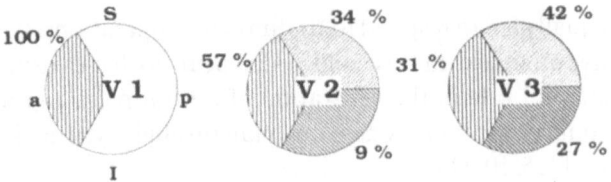

Fig. 3. Location of the conflict in patients with a single and clear arterial offending vessel and pain in one division only (i.e., 56 patients)

and the site of the offending vessel in relation to the root (i.e. anterior, superior, posterior and inferior quadrants) in the 148 patients of the series in whom there was only one conflicting arterial vessel.

Figure 3 shows the location of the conflicts in patients having a single and clear arterial offending vessel and pain in one division only (i.e. 56 patients). Pain in V1 was present only when the conflict was medial or medio-superior (100%). The incidence of pain in V2 was 57%, 34% and 9% with conflicts in the medial, superior and inferior aspect, respectively. The incidence of pain in V3 corresponded to conflicts in the medial (31%), the superior (42%), and the inferior (27%) aspects.

Discussion

Any conclusion on the role of neuro-vascular conflict in the genesis of idiopathic trigeminal neuralgia must take into account anatomical studies performed

in humans without a history of TN. In 50 autopsy studies of the trigeminal nerve, Hardy and Rhoton [5] found that in 26 cases there was a contact with the SCA with a few fascicles indented or distorted. Haines *et al.* [6] found a neurovascular contact (SCA) in 14 of their 40 specimens, but in ten the vessel did not distort the nerve.

In a personal series of 54 patients from the past 20 years who underwent a juxtapontine section of pars major of the trigeminal nerve for pain in the face due to malignant disease, we never found any significant vascular conflict. Although the mean age of this group was ten years younger (50 yrs) than in the series treated by MVD (60 yrs), it might serve as a control group. These anatomical data support the idea that an offending vessel plays a role in the genesis of idiopathic TN.

When reviewing our surgical data, we were surprised by the high frequency of accompanying abnormalities in close vicinity to the nerve, and the finding of a global atrophy of the root (67% of the cases) independent of a neuro-vascular conflict. These lesions which are not directly related to an offending vessel should be taken into account when discussing the pathogenesis of the disease. The presence of an offending vessel does not seem to be the sole explanation.

References

1. Dandy WE (1934) Concerning the cause of trigeminal neuralgia. Am J Surg 24: 447–455
2. Gardner WJ (1962) Concerning the mechanism of trigeminal neuralgia and hemifacial spasm. J Neurosurg 19: 947–958
3. Jannetta PJ (1976) Microsurgical approach to the trigeminal nerve for tic douloureux. Prog Neurol Surg 7: 180–200
4. Sindou M, Amrani F, Mertens P (1990) Décompression vasculaire microchirurgicale pour névralgie du trijumeau. Comparaison de différentes modalités techniques. Neuro-chirurgie 36: 16–26
5. Hardy DG, Rhoton A Jr (1978) Microsurgical relationships of the superior cerebellar artery and the trigeminal nerve. J Neurosurg 49: 669–678
6. Haines SJ, Jannetta PJ, Zorub DS (1980) Microvascular relations of the trigeminal nerve. An anatomical study with clinical correlation. J Neurosurg 52: 381–386

Correspondence: Marc Sindou, M.D., D. Sc., Department of Neurosurgery, Hôpital Neurologique Pierre Wertheimer, 59 Boulevard Pinel, F-69003 Lyon, France.

Acta Neurochir (1995) [Suppl] 64: 128–131

The Duke Experience with the Nucleus Caudalis DREZ Operation

J. P. Gorecki and **B. S. Nashold**

Department of Surgery, Division of Neurology, Duke University Medical Center, Durham, NC, U.S.A.

Summary

The nucleus caudalis DREZ operation has been performed in three phases at Duke. Between 1982 and 1988 radiofrequency (RF) lesions were made in the trigeminal nucleus extending from the C2 root to the obex using a straight electrode. Complications include ipsilateral arm ataxia due to spinocerebellar tract injury and ipsilateral lower limb weakness from the pyramidal tract. The former occurred at least transiently in 90% of cases. The electrode employed from 1988 to 1989 had proximal insulation protecting the spinocerebellar tract. Since 1989 a ninety degree bend has been added to the electrode to allow better placement. Two electrodes are used to accommodate the shape of the caudalis nucleus. A total of 101 procedures have been performed. The newest electrodes were used in 46 procedures. Ataxia is recognized in 39%. Overall pain relief was excellent in 34% and good in 40%. In post herpetic neuralgia 71% enjoyed excellent or good relief. Indications include post herpetic neuralgia, deafferentation pain (anaesthesia dolorosa, post-tic dysesthesia, stroke, MS, gasserian tumour, Gamma Knife radiation injury), facial trauma/surgery, atypical facial pain, and migraine/cluster headache. A study to compare this operation to deep brain stimulation prospectively for the above indications has been initiated.

Keywords: Facial pain; DREZ; trigeminal nucleus.

Introduction

The aim of DREZ lesioning is to affect the second order neurons in the sensory pathway. There is a growing experience with this procedure in the spinal cord and it has proven effective at relieving pain of central origin, particularly secondary to root avulsion [13–15]. The nucleus caudalis harbors second order neurons subserving pain, temperature and crude touch from the ispilateral face [18]. Trigeminal neuralgia responds well to medical management, microvascular decompression, peripheral or retro-gasserian lesions, however some patients develop post denervation dysesthesia or fail to improve. Patients with post herpetic neuralgia, atypical facial pain, pain following traumatic injury, or central pain secondary to stroke equally fail to respond to traditional tic operations. It is hypothesized that second order neurons have a substantial role in the generation of central pain, and in some instances may generate pain through hyperactivity [1, 5, 10, 11].

Sjoquist suggested and performed open trigeminal tractotomy which was reported in 1938 [20]. This was refined with stereotactic trigeminal tractotomy [6] and trigeminal nucleotomy [9,19]. Nashold extended the principals of DREZ to the trigeminal nucleus caudalis which was first reported in 1984 [16]. The purpose of this report is to review the Duke experience with this procedure over the last decade and describe the current surgical technique.

Materials and Methods

This is retrospective review of patients treated for facial pain with nucleus caudalis DREZ who failed maximum medical and surgical treatment. All patients were evaluated in a multidiciplinary pain clinic prior to referral for DREZ.

The procedure is performed under general anaesthesia with the patient prone and the head supported in pins. A small craniectomy is performed ipsilateral to the pain extending across midline to allow exposure of the obex, and reaching to the lateral margin of the foramen magnum. The arch of C1 is removed on this side. Magnification is employed. Landmarks are the obex, the rootlets of C2, motor rootlets of C1 and the accessory branches of the eleventh cranial nerve. A single row of lesions is made beginning at the rootlets of C2 in line with the dorsal root entry zone extending cephelad. Lesions are made with thermocouple temperature control for 20 seconds at 75 degrees Centigrade, one mm apart. Approximately 10 lesions are made to the level of C1 using the smaller angled electrode. The remaining six to ten lesions are made with the longer angled electrode extending up to the level of the obex. The final lesion is often in tissue which is notably softer to penetration and there is a noticeable decrease in impedance. The line of the lesions is carried in a gentle curve extending slightly lateral and care is taken to keep the lesions medial to the exit of the motor roots of eleven. The cerebellum and sometimes the PICA is gently elevated for the upper exposure (see Fig. 1). Patients are treated with high dose steroids.

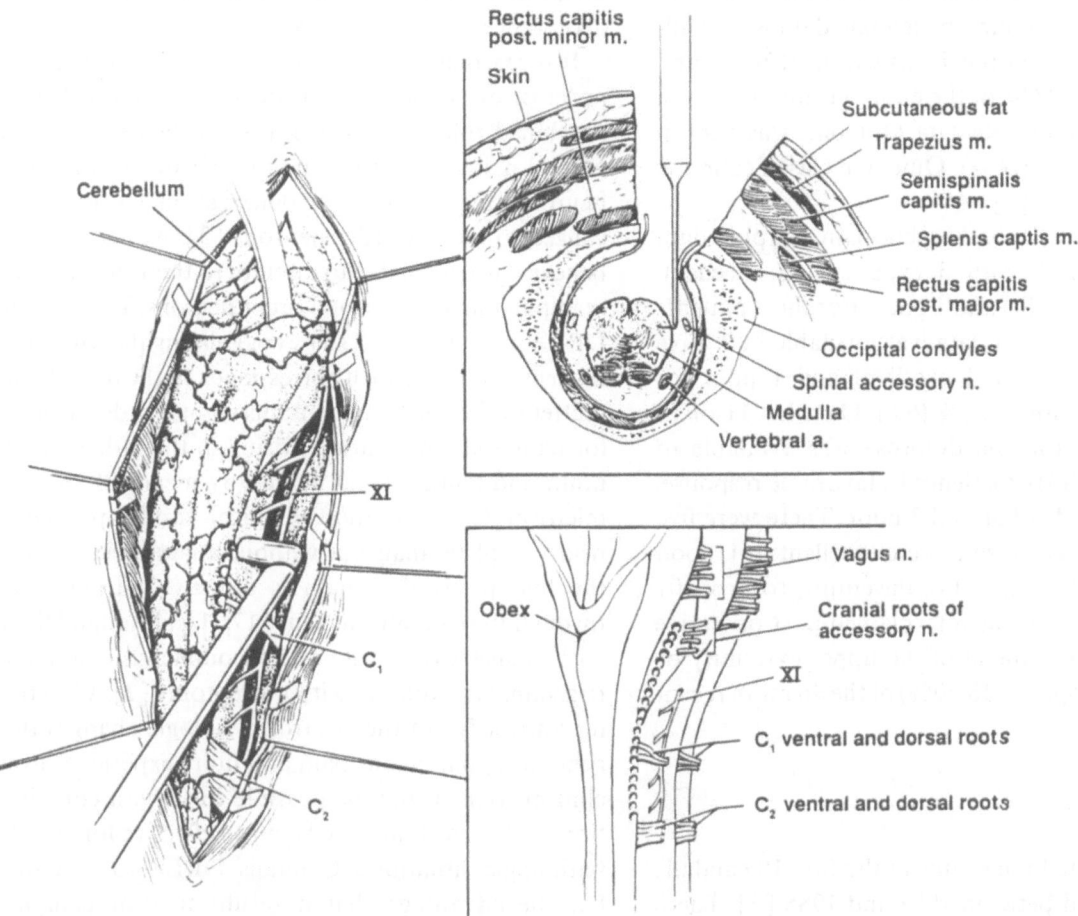

Fig. 1. Operative site. With permission of Nashold BS Jr [17]

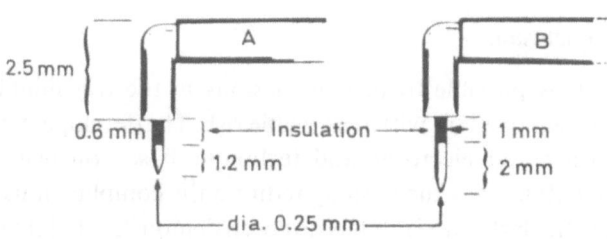

Fig. 2. El-Naggar/Nashold electrode

This technique is based on the results of an anatomical study on human autopsy cases by Drs. MM Abdulhak and J Ovelmen-Levitt. The nucleus caudalis merges with the cervical dorsal horn at C2 and extends 20 mm to the trigeminal nucleus intermedius, cephalad. The shape of the nucleus varies, being oval in cross section at the C2 level and measuring 0.8 mm × 1.5 mm. Near the obex the nucleus caudalis is round, with a cross section diameter of 2.0 mm. Superficial to the nucleus lies the spinal tract of the trigeminal nerve and the dorsal spinocerebellar tract which overlies the nucleus 3.0 mm bellow the obex. The pyramidal fibers cross and move from a ventral position to the dorsolateral quadrant of the cord immediately adjacent to the caudalis nucleus from a level between 5 mm and 12 mm caudal to the obex. The pyramidal tract is vulnerable to lesions that are made too deep particularly at the lower extent of the exposure. The El-Naggar/Nashold electrodes were introduced into use January 1990 [17]. The electrodes have a right angle to allow more accurate placement from the dorsal exposure. The shorter electrode measures 1.8 mm with the proximal 0.6 mm insulated and is used for the more caudal lesions to protect the pyramidal tract. The longer electrode is used more cephelad to more completely lesion the larger portion of the nucleus and it measures 3.0 mm with the proximal 1.0 mm insulated. A straight electrode with a 3.0 mm tip and proximal insulation over 1.0 mm to protect the spinocerebellar tract was introduced in 1988 [24]. The original procedure was begun in 1982 with a straight 2.5 mm electrode [2].

Results

A total of 101 nucleus caudalis DREZ lesionings have been performed at Duke to date, and 46 procedures have been performed with the El-Naggar/Nashold electrode. Fourteen patients had anaesthesia dolorosa following multiple procedures for trigeminal neuralgia, 8 had atypical facial pain, 8 had post herpetic pain, 5 stroke, 4 prior facial trauma or dental/nasal surgery, 4 had headache and one each had MS, trigeminal tumour, or Gamma Knife radiation for acoustic tumour. Mean follow up is 3 months with minimum follow up of one month. Current follow up data collection is ongoing. Inadequate data was available for 11 cases. Pain relief was excellent in 12 and good in 14, that is 56% of the entire group or 74% of the 35 cases with complete data. Outcome was poor in 6 and fair in 3 cases (26%). Outcome classification is previously described [2].

For the 8 cases with post herpetic pain complete data was available on 7 with 3 experiencing excellent, 2 good, 1 fair pain relief and 1 poor. For the 8 patients with atypical facial pain data was available in 7. Five were classified as good, 1 excellent and 1 poor for a favorable response of 75% (6/8). Data for 11 of 14 patients with anaesthesia dolorosa was available to evaluate. 50% (7/14) experienced a favorable response, 4 excellent, 3 good, 1 fair and 3 poor. There were five patients with stroke. Results were excellent in 1, good in 3 and not available in 1. For the entire group of 101 procedures there was one unexplained post operative sudden death. New ataxia of the upper extremity or gait ataxia developed in 18(39%) of the 46 more recent procedures.

Discussion

Bernard reported the results of the first 18 caudalis DREZ performed between 1982 and 1985 [2]. There was an incidence of at least transient upper extremity ataxia of 90% [17]. On immediate post operative evaluation 17 out of the 18 patients reported good or excellent outcome, while on subsequent follow up this fell to 11(58%). With a mean follow up of 9.8 months 71% of 7 patients with post herpetic pain reported a favorable response.

In 1994 21 cases operated with El-Naggar/Nashold electrodes between January 1990 and September 1992 were reviewed [17]. Seven patients developed postoperative ataxia, a substantial improvement from earlier reports. Pain relief was excellent in 48% and good in 5%. Four out of five patients with post herpetic neuralgia experienced excellent outcome.

There are several pathophysiologic mechanisms which result in pain in the face. Post herpetic pain is believed to represent a form of central pain with abnormalities in the ganglion and nucleus. Deafferentation pain is difficult to treat and many interventions have been proposed. A partial list includes, peripheral trigeminal lesions, ganglion lesions, root lesions, trigeminal tractotomy, trigeminal nucleotomy, undermining the skin [23], sympathectomy [8], thalamotomy [12], excision of sensory cortex and even cingulotomy [22].

So far none of these procedures has attained universal acceptance.

It is reasonable to theorize that second order, third order or even cortical neurons have an individual or combined role in the generation of the experience of central pain. It is probably not as simple as isolated neuronal hyperactivity or denervation sensitivity localized to second order neurons [3]. Nonetheless we believe it is reasonable to attempt to treat such pain by manipulating these second order neurons. The caudalis DREZ procedure does so by coagulating the trigeminal nucleus caudalis which corresponds to the dorsal gray matter of the cord and contains second order neurons for pain and temperature arising in the fifth, seventh, ninth and tenth cranial nerves, by means of an open microsurgical procedure. Clinical results producing near complete analgesia without anaesthesia support the hypothesis that pain is selectively transmitted through the caudalis nucleus [17,21], although Hitchcock suggests otherwise and recommends combining trigeminal tractotomy with nucleotomy [7]. With better knowledge of the electrophysiologic changes that generate spontaneous pain and more experience from clinical trials it may be possible to suggest combinations of lesions that may be more effective for specific pathologic situations. Caudalis DREZ as described has the advantage that it results in near complete destruction of the entire length of the nucleus as opposed to a focal area in stereotactic nucleotomy [4].

Conclusion

It is possible to produce lesions in the trigeminal nucleus caudalis with reasonable risk. The development of a newer electrode and technique based on neuroanatomy has successfully reduced the complications, particularly ataxia. It has not been demonstra- ted that the newer method improves the extent of nucleus destruction. Outcomes in terms of pain relief have remained at least as good, with favorable pain reduction in 73% at short term follow up. We anticipate some loss of good effect over time [2]. Results to date encourage further investigation of the technique and the initiation of a prospective trial to compare caudalis DREZ and thalamic stimulation in post herpetic and deafferentation pain of the face.

Acknowledgements

The authors would like to acknowledge the contributions of Linda L. Rubin RN MPH and Robbin Sharpe.

References

1. Anderson LS, Black RG, Abraham J, Ward Jr AA (1971) Neuronal hyperactivity in experimental trigeminal deafferentation. J Neurosurg 35: 444–452
2. Bernard EJ Jr, Nashold BS Jr, Caputi F, Moossy JJ Jr (1987) Nucleus caudalis DREZ for facial pain. Br J Neurosurg 1: 81–92
3. Dubner R, Bennett GJ (1983) Spinal and trigeminal mechanics of nociception. Ann Rev Neurosci 6: 318–418
4. El-Naggar Amr O, Nashold BS Jr (1995) Nucleus caudalis DREZ lesions for relief of intractable facial pain. In: Wilkins (ed) Textbook of neurosurgery
5. Hirayama T, Dostrovsky JO, Gorecki JP, Tasker RR, Lenz FA (1989) Recordings of abnormal activity in patients with deafferentation and central pain. Stereotac Funct Neurosurg 52: 120–6
6. Hitchcock ER, Schvarcz JR (1972) Stereotaxic trigeminal tractotomy for post herpetic facial pain. J Neurosurg 37: 412–417
7. Hitchcock ER, Teixeira MJ (1987) Pontine stereotactic surgery and facial nociception. Neurol Res 9: 113–117
8. Hyndman OR (1939) Post herpetic neuralgia in the distribution of the cranial nerves. Arch Neurol Psychiat 42: 224–232
9. Kerr FWL (1966) Spinal V nucleolysis for intractable craniofacial pain. Surg Forum 17: 419–421
10. Loeser JD, Ward Jr AA (1967) Some effects of deafferentation on neurons of the cat spinal cord. Arch Neurol 17: 629–636
11. Loeser JD, Ward Jr AA, White Jr LE (1968) Chronic deafferentation of human spinal cord neurons. J Neurosurg 29: 48–50
12. Mark VH, Ervin FR (1969) Stereotactic surgery for relief of pain. In: White JC, Sweet WH (eds) Pain and the neurosurgeon. Thomas, Springfield, III, pp 843–887
13. Nashold BS Jr, Urban B, Zorub DS (1976) Phantom pain relief by focal destruction of the substantia gelatinosa of rolando. Adv Pain Res Ther 1: 959–963
14. Nashold BS Jr, Ostdhl RH (1979) Dorsal root entry zone lesions for pain relief. J Neurosurg 51: 69
15. Nashold BS Jr, Ostdhl RH, Bullitt E (1983) Dorsal root entry zone lesions: a new neurosurgical therapy for deafferentation pain. Adv Pain Res Ther 5: 739–750
16. Nashold BS Jr, Caputi F, Bernard E (1984) Trigeminal DREZ: caudalis nuclear lesions for relief of facial pain. Neurosurgery 19: 150
17. Nashold BS Jr, El-Naggar Amr O, Ovelmen-Levitt J, Muwaffak A (1994) A new design of radiofrequency lesion electrodes for use in the caudalis nucleus DREZ operation. J Neurosurg 80: 1116–1120
18. Olszewski J (1950) On the anatomical and functional organization of the spinal trigeminal nucleus. J Comp Neurol 92: 401–413
19. Schvarcz JR (1977) Postherpetic craniofacial dysaesthesiae: their management with stereotactic trigeminal nucleotomy. Acta Neurochir (Wien) 38: 65–72
20. Sjoqvist O (1938) Studies on pain conduction in the trigeminal nerves; a contribution to the surgical treatment of facial pain. Acta Psychiatr Neurol [Suppl] 17: 1–139
21. Sampson JH, Nashold BS Jr (1992) Facial pain due to vascular lesions of the brain stem relieved by dorsal root entry zone lesions in the nucleus caudalis. J Neurosurg 77: 473–475
22. Sugar O, Bucy PC (1951) postherpetic trigeminal neuralgia. Arch Neurol Psychiat 65: 131–145
23. Tindall GT, Odum GL, Vieth RG (1962) Surgical treatment of postherpetic neuralgia: results of skin undermining and excision in 14 patients. Arch Neurol 7: 423–426
24. Young JN, Nashold BS Jr, Cosman ER (1989) A new insulated caudalis nucleus DREZ electrode. J Neurosurg 70: 283–28

Correspondence: J.P. Gorecki M.D., FRCS (C), Department of Surgery, Duke University Medical Center, Durham, NC 27710, U.S.A.

Acta Neurochir (1995) [Suppl] 64: 132–135

Cortical Stimulation for Central Neuropathic Pain: 3-D Surface MRI for Easy Determination of the Motor Cortex

P. Herregodts, T. Stadnik[1], F. De Ridder[1], and J. D'Haens

Departments of Neurosurgery and [1] Radiology, University Hospital Free University of Brussels (AZ-VUB), Brussels, Belgium

Summary

Motor cortex electric stimulation has been reported to be effective for the treatment of central post-stroke pain and trigeminal neuropathic pain. Five patients with pain due to injury of the trigeminal nerve and with abnormalities of facial sensibility, as well as two patients suffering of a post-stroke thalamic pain, were subjected to stimulation applied epidurally on the motor cortex. Quadripolar electrodes were implanted under local anaesthesia and the precise location of the motor cortex was determined on three-dimentional surface MRI the day prior to surgery. In our experience, correct topographic localization of the electrode on the motor cortex seems to be crucial to obtain pain reduction.

Keywords: Neuropathic pain; motor cortex; neurostimulation; magnetic resonance imaging.

Introduction

Motor cortex stimulation has been proposed as a treatment modality in deafferentation pain (Namba and Nishimoto 1988). Tsubokawa *et al.* (1991, 1993) reported beneficial effect of epidurally applied motor cortex stimulation in patients with therapy-resistant central pain after thalamic stroke. Meyerson *et al.* (1993) further extended the indications of this technique by reporting pain relief in trigeminal neuropathic pain.

One of the main technical problems arising when applying this technique is the precise determination of the gyrus precentralis for correct epidural implantation of the stimulating electrodes. Previous reports utilizing conventional methods for determination of the motor cortex, such as bone landmarks of the skull (Delmas and Pertuiset, 1959), stereotactic atlasses (Talairach *et al.*, 1957) as well as somatosensory evoked potential recording (Wood *et al.*, 1988) seem to have some drawbacks as to the precision in determining the motor cortex and stimulation site, as was discussed

previously (Meyerson *et al.*, 1993). We describe here a non-invasive 3-D MRI method for brain surface reconstruction and cortex determination, enabling fast localization of the motor cortex. The MRI procedure is performed prior to surgery and accurate electrode position is confirmed by functional stimulation testing per-operatively.

Materials and Methods

Patient Population

A total of 7 patients with central pain were selected for motor cortex stimulation between January 1993 and October 1994. Five of the patients had trigeminal neuropathy (in 4 of these due to injury of the nerve after surgical trigeminal rhizotomy or decompression) and 2 patients had a post-stroke central pain (a thalamic hemorrhage and a small thalamic infarct, both with severe thalamic pain predominantly located in the face and upper extremity). All patients had failed to respond to pharmacological therapy (analgesics, antiepileptics and psychotropic drugs), as well to TENS or spinal cord stimulation applied cervically in some patients. In one patient with thalamic pain due to a small thalamic ischemic lesion deep brain stimulation in the VPM failed to give pain relief. There was no history of prior epilepsy in any of our patients. They all had, prior to implantation, EEG examination for the exclusion of epileptic activity. A routine diagnostic MRI procedure in axial and sagittal planes was performed prior to implantation to exclude major morphological alterations of the cerebral cortex.

MRI Procedure for Determination of the Motor Cortex

The day before surgery, the presumed region of motor cortex is covered with an external reference grid composed of 9 tubings intersecting at 90°, spread by 1 cm and filled with a $CuSO_4$ solution. This external reference is firmly attached to the skin of the patient with tape and the location of the grid is also drawn on the patients skin (Fig. 1). A 3-D FLASH 40/6/1/40 (TR/TE/Number of excitations/Flip angle) sequence is then performed in the sagittal plane after intravenous Gd-DOTA injection on 1T Magnetom (Siemens, Germany). The slice thickness is 2.0 mm, the number of excitations 1,

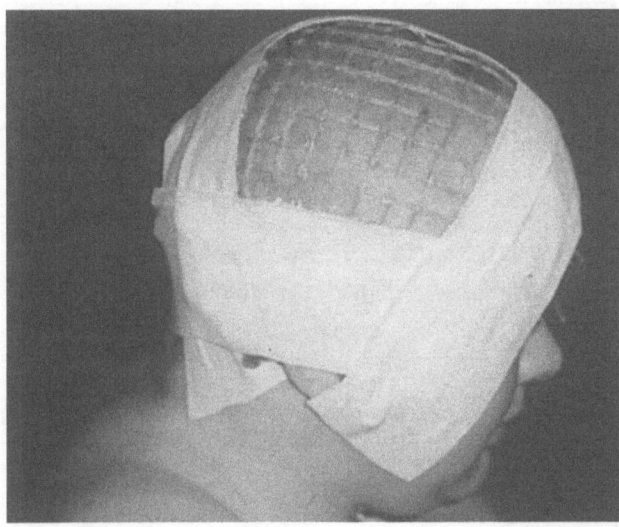

Fig. 1. Lateral view of the external reference grid attached to the skull of the patient

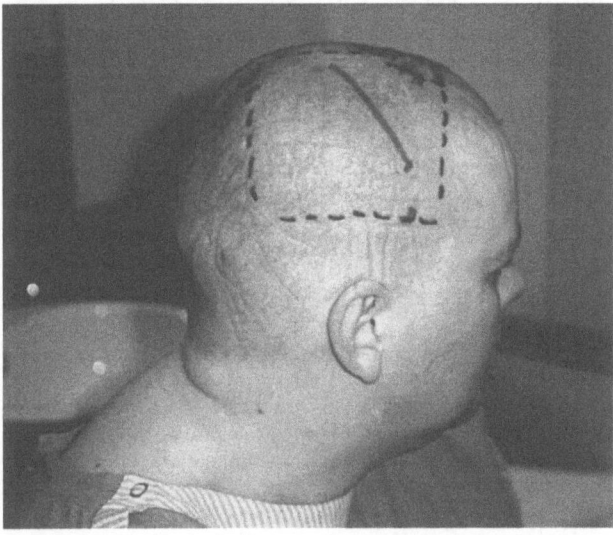

Fig. 3. A line is drawn on the skin corresponding to the localization of the motor cortex

the matrix 256 × 256 and the number of slices 128. This basic 3-D MRI data acquisition time is 18 minutes. On surface reconstruction based on the 3-D Tubo-FLASH (Ebeling *et al*, 1989) the normal gyri and sulci of the cortex can be easily recognized. A series of curved 3-D surface reconstuctions (using a 3-D display program) is then performed, plotting the external grid (sticked to the skin of the head) onto the 3-D reconstructed cortex (Fig. 2). The g. precentralis is then identified and a line corresponding to the site of the motor cortex is drawn on the patients skin (Fig. 3).

Surgical Procedure

Under local anaesthesia a burr hole was made 4 cm lateral to the midline at the site of the previously drawn line on the patient's skin. Once the dura was exposed, it was detached from the bone in the direction of the same line and a 4-polar electrode strip (Resume, Medtronic Inc.) was introduced epidurally. Test stimulation via a percutaneous extension lead was then performed peroperatively. First, low frequency (1–2 Hz) high intensity stimulation (4–8 V, 0.8 ms

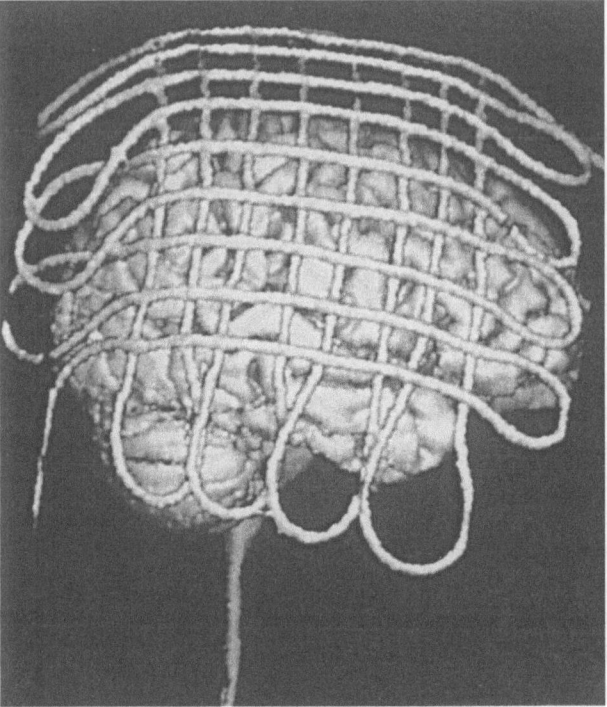

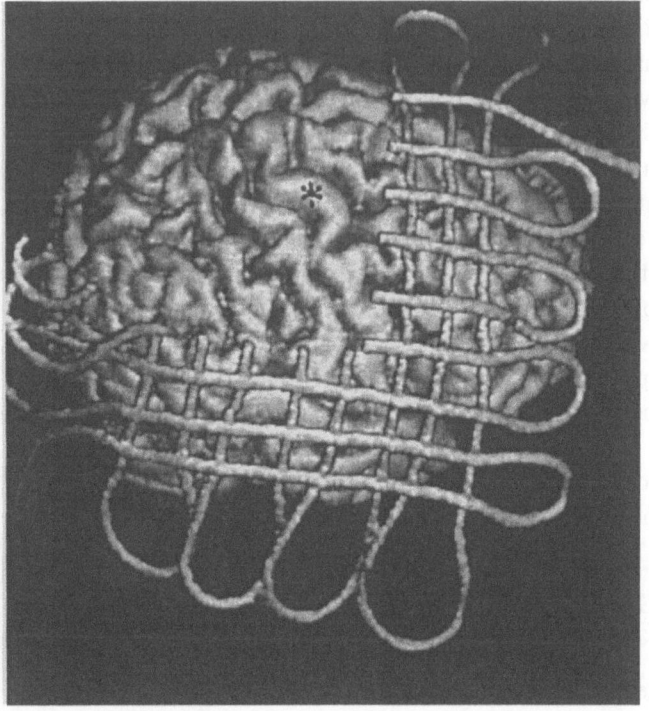

Fig. 2. Surface reconstruction (lateral view) based on 3-D FLASH (without Gd injection) showing the different cortical gyri and sulci with the grid superimposed onto the cortical structures. The g. precentralis is identified (*): right side

duration) was applied till muscle switches were observed in the face or upper extremity corresponding to the painful area. If necessary, the quadripolar electrode position was corrected to cover the appropiate cortex area. Secondly, stimulation was performed at high frequency (50–75 Hz) and with an intensity of about 2/3 or the motor threshold (0.2 ms duration). The majority of patients reported a sensation of light tingling or mild vibration projected in the same area of distribution as their pain when the electrode was considered correctly positioned.

Stimulation Procedure

A few days after electrode implantation the patient was discharged and continued trial stimulation via percutaneous extension leads using a screening stimulator. This continued for a period of a least 4 weeks. Patients were instructed to stimulate at the parameters determined during high-frequency test stimulation per-operatively for a period of one hour 6 times a day. After this period pain was reassessed by using the VAS (visual analogue scale) and compared to prestimulatory values. If a mean pain relief of 50% or more was achieved, the patient was considered suitable for continued stimulation using a completely implantable system (Itrel II, Medtronic Inc.).

Results

The results are summarized in Table 1. In all patients implanted the motor cortex could easily be identified using 3-D surface reconstruction MRI. Test stimulation during implantation could elicit local muscle twitches in the painful area (in most patients in the face, in one patient also the upper leg contralateral to the stimulation side). The best results were obtained in patients with trigeminal neuropathic pain: 4 of the 5 patients reported a consistent pain relief of 50% or more. In 1 patient with anaesthesia dolorosa pain relief was only of 20% after 6 weeks of trial stimulation. Since the patient insisted to continue the treatment, she had an implanted system. Unfortunately, 8 months after electrode implantation the pain relieving effect ceased. Of the two patients suffering from post-stroke thalamic pain, one with predominantly facial and neck pain reported a 50% pain relief. The second patient with pain mainly located in the upper extremity and ipsilateral chest had no pain relief during the trial period.

3-D MRI control after electrode implantation performed in 3 patients confirmed the correct positioning of the electrode on the motor cortex (Fig. 4).

Discussion

The major issue of this paper is to describe the technical features and the application of 3-D surface MRI. In our experience, this technique is of great help to correctly position the electrodes onto the motor

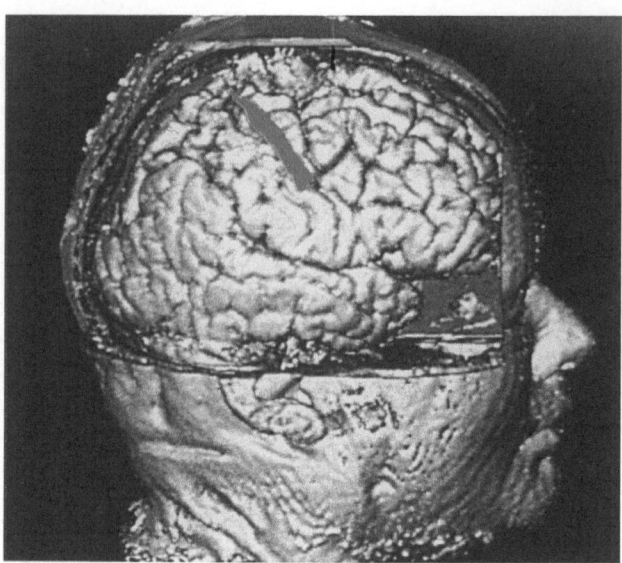

Fig. 4. 3-D surface MRI control after implantation confirming the correct position of the quadripolar electrode onto the motor cortex

Table 1. *Patients Treated with Motor Cortex Stimulation Results*

Sex, age (yrs)	Diagnosis	Follow-up (months)	Pain-relief (% red. VAS)
F, 40	anaesth. dolorosa	22	70
F, 43	anaesth. dolorosa	20	100
M, 52	trig. neuralgia	11	50
F, 58	anaesth. dolorosa	10	50
F, 65	anaesth. dolorosa	13	20 - >0
F, 41	thal. hemorrhage	9	50
M, 62	thal. infarct	4	0

cortex. The results in the 7 patients treated by this technique (Table 1) are comparable with data previously reported by other groups (Tsubokawa *et al.*, 1993; Meyerson *et al.*, 1993). The correct positioning of the electrodes onto the motor cortex seems to be crucial for the production of pain relief. Since the electrodes are postioned epidurally and the location is checked by per-operative test stimulation, it is advantageous to perform the implantation procedure under local anaesthesia. During the operation it is not possible to identify the different cortical gyri and therefore, we have developed a non-invasive 3-D surface MRI method to determine the location of the motor cortex with a precision of 1–2 mm (corresponding to the spatial distortion of the MRI). This technique is independent of bony landmarks and stereotactic atlasses. Most MRI machines are presently equipped with a 3-D software programme and many centres have acquired experience in this field (e.g. Ehricke *et al.*, 1992; van Veelen 1994). In our experience the MRI procedure with 3-D surface reconstruction takes about 1 hour and can be performed on an out-patient basis.

References

1. Delmas A, Pertuiset B (1959) Topométrie cranio-encéphalique chez l'Homme. Masson, Paris
2. Ebeling U, Steinmetz H, Huang Y, Kahn T (1989) Topography of the inferior precentral sulcus in MR imaging. AJNR 10: 937–942
3. Enricke HH, Daiber G, Sonntag R, Straβer W, Lochner M, Schad LR, Lorenz WJ (1992) Interactive 3D graphics workstations in stereotaxy: clinical requirements, algorithms and solutions. Proc. visualization in biomediacal cumputing, Chapel Hill, NC, SPIE, 1808, pp 548–558
4. Meyerson BA, Lindblom U, Linderoth B, Lind G, Herregodts P (1993) Motor cortex stimulation as treatment of trigeminal neuropathic pain. Acta Neurochir (Wien) 58 [Suppl]: 150–153
5. Namba S, Nishimoto A (1988) Stimulation of internal capsule, thalamic sensory nucleus (VPM) and cerebral cortex inhibited deafferentation hyperactivity provoked after Gasserian ganglionectomy in cat. Acta Neurochir (Wien) 42 [Suppl]: 243–247
6. Talairach J, David M, Tournous P, Corredor H, Kvasina T (1957) Atlas d'anatomie stéréotaxique. Masson, Paris
7. Tsubokawa T, Katayama Y, Yamamoto T, Hirayama T, Koyama S (1991) Chronic motor cortex stimulation for the treatment of central pain. Acta Neurochir (Wien) 52 [Suppl]: 137–139
8. Tsubokawa T, Katayma Y, Yamamoto T, Hirayama T, Koyama S (1993) Chronic motor cortex stimulation in patients with thalamic pain. J Neurosurg 78: 393–401
9. van Veelen KWM (1994) Three-dimentsional visualisation of depth electroencephalography. In: Kupers R (ed) A look into the seeds of time. Leuven University Press, Leuven, pp 167–172
10. Wood CC, Spencer DD, Alison T (1988) Localization of human sensorimotor cortex during surgery by cortical surface recording of somatosensory evoked potentials. J Neurosurg 68: 99–111

Correspondece: Patrick Herregodts, M.D., Ph.D., Department of Neurosurgery, University Hospital Free University Brussels (AZ-VUB), Laarbeeklaan 101, B-1090 Brussels, Belgium.

Acta Neurochir (1995) [Suppl] 64: 136–138

A New Approach to the Control of Central Deafferentation Pain—Spinal Intrathecal Baclofen

T. Taira, H. Kawamura, T. Tanikawa, H. Kawabatake, H. Iseki, A. Ueda, and **K. Takakura**

Department of Neurosurgery, Neurological Institute, Tokyo Women's Medical College, Tokyo, Japan

Summary

We investigated the short-term effects of an intrathecal bolus injection of baclofen on central pain due to stroke or spinal cord injury. Pain relief was obtained in 64% of the patients. The effects developed 1–2 hours after the injection and continued for 10–24 hours. Both spinal segmental and supraspinal mechanisms may be involved in the production of baclofen-analgesia.

Keywords: Baclofen; central pain; GABA; pain relief.

Introduction

Central pain following stroke or spinal cord injury is notoriously difficult to control. In a patient who was given spinal intrathecal baclofen for post-stroke spasticity, we incidentally found that dysesthetic pain in the extremities was substantially relieved and the effect persisted for about 24 hours. Beside an antispastic effect, baclofen, which is an agonist of gamma aminobutylic acid (GABA), has in experimental animals antinociceptive effects that are not reversed by naloxone [2,4,6,12,13]. Furthermore, intrathecal administration of baclofen reduces touch evoked responses in animal models of allodynia [8]. It has also been reported that intrathecal baclofen may have favourable effects on central pain due to spinal lesions [6]. These data prompted us to investigate the short-term effects of an intrathecal bolus injection of baclofen in patients with central pain.

Materials and Methods

Eight patients with post-stroke central pain and six patients with central pain due to spinal cord injury were studied. Clinical data on the patients are summarized in Table 1. All the patients had been extensively treated with oral analgesic medication but with unsatisfactory pain relief. Case 1 had undergone both deep brain stimulation and motor cortex stimulation and Case 2 had had deep brain

stimulation, but in both patients the effects were modest and transient. The diagnosis of central pain was made on the basis of clinical examination and was further supported by the finding that the pain was insensitive to intravenous morphine. Radiological findings with CT or MRI were consistent with the neurological signs in all the patients. No patient had tried oral baclofen before the study.

A bolus of intrathecal baclofen (50–100 µg) was administered once a day via a lumbar puncture at the L3–4 level. The baclofen solution for intrathecal use was supplied by Ciba Geigy Corporation (Basel, Switzerland). The patients were asked to assess their pain hourly using a 10-grade score (0: no pain, 10: pain of pre-treatment level). We did not tell the patients about the expected time-course of a possible pain relief. The injection was repeated 3–5 times over a week. Normal saline was used once in each patient to exclude a placebo effect. We explained thoroughly to the patients and their family the experimental nature of the study and the possible risks, and obtained informed consent. The procedure was approved by the Ethics Committee of Tokyo Women's Medical College.

Results

Six patients (75%) with post-stroke pain and three patients (50%) with pain due to spinal cord injury reported a substantial pain relief which they had previously never experienced. The effect generally appeared 1–2 hours after the injection and persisted for 10–24 hours. A typical example of the time course of pain relief is shown in Table 2. Allodynia and hyperalgesia, if present, were alleviated as well. Pinprick and light touch sensations did not change in the non-affected extremities. Placebo injections had no effect in any of the patients. In the post-stroke group, clinical and radiological findings in the responders did not differ from those in the non-responders. In the group of spinal cord injury, patients with signs of partial spinal lesions reported pain relief, whereas patients with complete lesions of the spinal cord did not benefit. The results are summarized in Table 1. In some patients the injection produced unsteady gait, urinary retention,

Table 1. *Summary of Patients*

Case		Site of lesion	Duration of symptoms (yrs)	Allodynia	Hyperalgesia	Best pain score
1.	57F	thalamus	0.5	Y	N	10/10
2.	47F	putamen	2	N	Y	4/10
3.	58F	putamen	2	N	N	10/10
4.	70M	thalamus	3	N	N	0/10
5.	60M	thalamus	5	Y	N	2/10
6.	71M	thalamus	12	Y	Y	3/10
7.	62M	corona radiata	15	N	N	2/10
8.	57F	pons	20	N	N	5/10
9.	46M	C3–4	8	Y	Y	2/10
10.	66M	Th4	10	Y	N	4/10
11.	58M	Th12	28	N	N	2/10
12.	65M	L1	8	N	N	10/10
13.	50M	Th12	10	N	N	10/10
14.	46M	C6	31	Y	N	10/10

Y yes, *N* no, *0/10* no pain, *10/10* pain of pretreatment level.

Table 2. *Typical Time Course of Pain Relief After Intrathecal Administration of Baclofen (Case 7). The original table was filled in by the patient*

Time	Arm pain	Leg pain	Event
9:00	10	10	
10:00	10	10	
11:00	10	10	50 µg baclofen,
12:00	5	5	intrathecal injection at 10:55
13:00	2–4	2	
14:00	2	2	
15:00	2	2	
16:00	1–2	2	
17:00	1–2	2	
18:00	1–2	2	
19:00	1–2	2	unsteady gait
20:00	—	—	sleeping
21:00	—	—	sleeping
22:00	—	—	sleeping
23:00	—	—	sleeping
24:00	—	—	sleeping
1:00	—	—	sleeping
2:00	—	—	sleeping
3:00	4	2	toilet
4:00	6	2	
5:00	6	2	
6:00	7	2	toilet
7:00	8	2	
8:00	8	2	
9:00	8	2	

and a mild headache, but these side-effects were transient.

Discussion

Contrary to general belief that interventions influencing spinal cord function may not control central pain of supraspinal origin, spinal intrathecal baclofen produced a substantial relief in 75% of the patients with post-stroke central pain. Obviously, this pain relief is not due to a placebo effect, because the response to the baclofen was consistently present in all these patients whereas the placebo injection was ineffective. Although the patients were not informed of the possible time course of pain relief, it was highly reproducible. Nine patients reported that no previous treatments had been so effective for their pain as intrathecal baclofen.

It is well documented that intrathecal baclofen may relieve muscle spasm pain, and this effect is generally believed to be secondary to the reduction of spasticity [3]. However, there are few clinical reports concerning pain relief with intrathecal baclofen. Magora *et al.* [7] successfully alleviated chronic low back pain with intrathecal baclofen. Because of the complex nature of chronic low back pain, it is difficult to judge whether the effect in that condition is primary or secondary. Herman *et al.* [5] reported that central pain caused by spinal cord lesions may be successfully controlled with intrathecal baclofen and obviously this is not the secondary effect. In their report also a patient with a C3 lesion experienced pain relief in the leg.

The mechanism of relief of central pain by intrathecal spinal baclofen is difficult to explain. As the reduction of pain started from the leg in the initial two patients tested, we first believed that the effect was primarily exerted on a segmental spinal level. However, such a mechanism could not account for the pain relief observed in case 4 and 5, in whom the pain alleviation was first experienced in the face and arm, respectively. The fact that patients with complete spinal cord transsection did not respond to intrathecal baclofen suggests that there may be an ascending pain control system which is triggered by a GABAergic system in the spinal cord.

Patients with central pain and allodynia/dysesthesia often protect themselves by covering their hands and feet with gloves and socks. Joint movements often evoke pain, while lying flat in bed reduces the pain. These clinical observations indicate that low threshold mechanical stimuli and activation of deep sensation play a role in the generation of central pain. Intathecal baclofen may modulate such impulses at a spinal cord level resulting in pain relief. Thus, both spinal and supraspinal mechanisms seem to be involved in the pain relief induced by intrathecal baclofen. It might be that the effect of intrathecal baclofen has mechanisms in common with spinal cord stimulation, which also may relieve both pain and spasticity.

Baclofen analgesia is not mediated via the involvement of the endogenous opioid system. Proudfit and Levy [10] showed that the neural structures rostral to the medulla and caudal to the midbrain are necessary for the analgesic effect of baclofen. These findings suggest that there is an ascending pain control system from the spinal cord to the pons which is not mediated by the opioid system. Because baclofen acts on GABA$_B$ receptor sites which are present in high concentration in the spinal dorsal horn [9], GABA may be the mediator of this pain control system. It has been reported that GABA is released by electrical spinal cord stimulation [11], which technique has been clinically used for pain relief since long and this finding further supports the importance of GABAergic systems in the generation of control of pain. A clear distinction should be made between noxious pain and central neuropathic pain. In an animal model of central pain induced by prostaglandin F2α, intrathecal baclofen suppressed touch evoked allodynic responses [8]. However, it has also been reported that a selective antagonist to the GABA$_B$ receptor, phaclofen, does not affect such allodynic responses [14]. The latter finding suggests that not GABAergic but other systems are involved in the suppression of allodynia. Because baclofen acts not only on GABA$_B$ receptors but it also agonizes glycine containing inhibitory neurons which are present in the dorsal horn [1], the effect of spinal intrathecal baclofen on allodynia might be mediated by these glycine neurons. If this is true, allodynia might be relieved by the administration of glycine, which readily passes the blood brain barrier.

The findings in the present study support the hypothesis of Yaks h and Yamamoto [14] that loss of dorsal horn GABA-and glycine- containing interneurons in human may result in allodynia or hyperesthesia. However, it is not known whether central pain of supraspinal origin is associated with a loss of interneurons in the spinal dorsal horn. This pilot study indicates that a controlled clinical trial of continuous baclofen infusion in patients with central pain of supraspinal origin is warranted. Hopefully, such studies would also deepen our understanding of the mechanisms of central pain.

References

1. Blumenkopf B (1991) The general aspects of neuropharmacology of dorsal horn function. Adv Pain Res Ther 151–176
2. Cutting DA, Jordan CC (1975) Alternative approach to analgesia: baclofen as a model compound. Br J Pharmac 54: 171–179
3. Gybels JM (1992) Indications for neurosurgical treatment of chronic pain. Acta Neurochir (Wien) 116: 171–175
4. Henry JL, Ben-Ari Y (1976) Action of the p-chlorophenyl derivative of GABA, Lioresal, on nociceptive and non-nociceptive units in the spinal cord of the cat. Brain Res 117: 540–544
5. Herman RM, D'Luzansky SD, Ippolito R (1992) Intrathecal baclofen suppresses central pain in patients with spinal lesions. Clin J Pain 8: 338–345
6. Levy RA, Proudfit HK (1977) The analgesic action of baclofen [β-(4-chlorophenyl)-γ-aminobutylic acid. J Pharmac Exp Ther 202: 437–445
7. Magora F, Magora A, Vatine JJ, Shochina M (1992) Clinical and electrophysiological effects of subarachnoid baclofen in low back pain. In: Hyodo M, Oyama T, Swerdlow (eds) The pain clinic IV, VSP Utrecht, pp 75–80
8. Minami T, Uda R, Horiguchi S, Ito S, Hyodo M, Hayaishi O (1992) Effects of clonidine and baclofen on prostaglandin F2α induced allodynia in conscious mice. Pain Res 7: 129–134
9. Price GW, Wilkin GP, Turnbull MJ, Bowery NG (1984) Are baclofen sensitive GABA$_B$ receptors present on the primary afferent terminals of the spinal cord? Nature 307: 71–73
10. Proudfit HK, Levy RA (1978) Delimitation of neuronal substrates necessary for the analgesic action of baclofen and morphine. Eur J Pharmac 47: 159–166
11. Stiller CO, O'Connors W, Linderoth B, Hammarstrom G, Brodin E (1992) PAG-microdialysis in the awake, freely moving rat during spinal cord stimulation: Release of GABA. Acta Neurochir (Wien) 117: 87 (abstract)
12. Wilson PR, Yaksh TL (1978) Baclofen is antinociceptive in the spinal intrathecal space of animals. Euro J Pharmac 51: 323–330
13. Yaksh TL, Reddy SVR (1987) Studies in the primate on the analgetic effects associated with intrathecal actions of opiates, α-adrenergic agonists and baclofen. Anesthesiology 54: 451–467
14. Yaksh TL, Yamamoto T (1991) Studies on the pharmacology of spinal systems underlying anomalous pain states. Adv Pain Res Ther 197–207

Correspondence: T. Taira, M.D., Department of Neurosurgery, Neurological Institute, Tokyo Women's Medical College, 8–1 Kawada-cho, Shinjukuku, Tokyo 162, Japan.

Index of Keywords

Area 24   69

Baclofen   136
Basal gangilia-thalamocortical motor
    circuit   5

Callosal grid localization   79
Cancer pain   88, 97
Central area   79
Central pain   136
c-fos   69
Chorea   5
Chromaffin cells   97
Cingulotomy   69
Clonidine   109
Cognitive process   83
Colloid cyst   59
Computer assisted neurosurgery   49
Computer assisted surgery   54
Computer assisted system   40
Computer modeling   119
Contour scanning   45
Cranial nerves   35
Craniopharyngioma   59
CT-guidance   88
Cystic brain tumour   59

3-D laser range imaging   45
3-D localisation   40
Dopa-induced dyskinesia   5
DREZ   128

EEG   74
Enkephalin   97
Epileptic foci   74
Evoked potentials   92

Facial pain   128
Failed back surgery   116

Failed back surgery syndrome   106
Flexion reflex   17
Foetal transplants   1
Frameless stereotaxy   45, 49, 54
Free flap   101
Functional MR imaging   83

GABA   136
Globus pallidus   30

Hypermetabolism   9

Image co-registration   45
Interactive image-guided surgery   54
Intrathecal baclofen   17, 26

Laser Doppler flowmetry   101

Magnetic resonance imaging   132
Magnetoencephalography   74
Medical imaging   40
Microcirculation   101
Microelectrode   9
Micro-vascular decompression   125
Morphine   109
Motor cortex   132
Multiple sclerosis   13

Neuroendoscopy   59
Neurogenic bladder   17
Neuronal graft   97
Neuropathic pain   132
Neurostimulation   132
Neurosurgical stereotaxy   40
Neuro-vascular conflict   125

Obsessive-compulsive-disorder   64
Opioids   97

Pain   30, 69, 83, 106, 109, 116, 125
Pain relief   136
Pallidotomy   9
Paresthesia coverage   119
Parkinsonian rigidity   5
Parkinson's disease   1, 30
Partial seizures   79
Perception threshold   119
Percutaneous cordotomy   92
Percutaneous procedures   88
PET   9
Pineal cyst   59
Posterior fossa surgery   35
Psychosurgery   30, 64, 83
Pump   109

Reoperation   106
Robotics   54

Spasticity   17, 26
Spinal cord position   119
Spinal cord stimulation   92, 101, 106,
    109, 116, 119
Spinal opoids   109
Stereotactic implants   1
Stereotactic neurosurgery   45, 59
Stereotactic surgery   13, 49
Stereotaxy   64, 88

Thalamotomy   13
Tremor   13
Trigeminal neuralgia   125
Trigeminal nucleus   128

Vasospasm   101
Viewing wand   49

WGA-HRP   69

Index of Keywords

SpringerNews

Wolfgang Koos, Bernd Richling (eds.)

Stereotactic Neuro-Radio-Surgery

Proceedings of the International Symposium
on Stereotactic Neuro-Radio-Surgery, Vienna 1992

1995. 101 partly coloured figures. VII, 119 pages.
Cloth DM 150,–, öS 1050,–
Reduced price for subscribers to "Acta Neurochirurgica":
Cloth DM 135,–, öS 945,–
ISBN 3-211-82657-2
Acta Neurochirurgica, Supplement 63

During the last few years stereotactic radiosurgery has become a partner of equal rank within the discipline of neurosurgery. Today it is regarded as being of the same importance as microsurgery and endovascular neurosurgery, branches which have also progressed rapidly in recent years. Breakthrough success, however, requires a combined effort of all partners involved.
The editors have brought together leading experts in the fields of neurosurgery, neuroradiology, neurology, neuropathology, neuroanatomy, radiation oncology, and biophysics to discuss indications and therapeutic strategies in the treatment of arteriovenous malformations and intracranial tumors and to find a common basis for their future work.

SpringerNeurosurgery

SpringerWienNewYork

P.O.Box 89, A-1201 Wien • New York, NY 10010, 175 Fifth Avenue
Heidelberger Platz 3, D-14197 Berlin • Tokyo 113, 3-13, Hongo 3-chome, Bunkyo-ku